On
Science

What are we to make of the attempts by recent science and scientists to find a Theory of Everything? Are there some things science just can't explain? Brian Ridley, a physicist, investigates these questions and others in this compelling exploration of both the scope and the limits of science.

Going back to the roots of scientific thinking in a world of magical ideas, Ridley argues that science shares more with magic than we are often led to believe. He also revisits Pythagoras's theory that the world should be understood through numbers, and explores the often overlooked relationship between science and mathematics. This is neatly linked to a fascinating discussion of relativity and quantum theory, reminding us of the many perspectives on offer within science. On Science closes with an important look at the often utopian scientific ideals of future societies, and returns to the problematic relation between science and sensibility that fuelled the Two Cultures controversies of the 1960s.

On Science is essential reading for all those interested in the way we think about and picture science, where it is now, and where it is going.

Brian Ridley is Professor of Physics at the University of Essex. He is best known for his work on semiconductor physics and for his best-selling book Time, Space and Things.

Thinking in Action

Series editors: **Simon Critchley**, University of Essex, and **Richard Kearney**, University College Dublin and Boston College.

Thinking in Action is a major new series that takes philosophy to its public. Each book in the series is written by a major international philosopher or thinker, engages with an important contemporary topic, and is clearly and accessibly written. The series informs and sharpens debate on issues as wide ranging as the Internet, religion, the problem of immigration and refugees and the way we think about science. Punchy, short, and stimulating, **Thinking in Action** is an indispensable starting point for anyone who wants to think seriously about major issues confronting us today.

B. K. RIDLEY

On
Science

Routledge
Taylor & Francis Group
LONDON AND NEW YORK

First published 2001
by Routledge
2 Park Square, Milton Park, Abingdon, Oxon OX14 4RN

Simultaneously published in the USA and Canada
by Routledge
711 Third Avenue, New York, NY 10017, USA

Routledge is an imprint of the Taylor & Francis Group, an informa business

Typeset in Joanna by
RefineCatch Ltd, Bungay, Suffolk

British Library Cataloguing in Publication Data
A catalogue record for this book is available from the British Library

Library of Congress Cataloging in Publication Data

Ridley, B. K.
   On science / Brian Ridley.
      p.   cm. – (Thinking in action)
   Includes bibliographical references and index.
      1. Science – Philosophy.   2. Mathematics.   I. Title.   II. Series.

   Q175 .R468 2001
   501 – dc21   00–045942

ISBN 978-0-415-24980-5 (pbk)

For
Aaron and Ann, Melissa and Anil

The motivation for writing this book has had several energizing strands. One is my perception of a widespread belief, often explicitly stated, that, given time, science will explain everything, including those eternal human mysteries that are traditionally the concern of religion and the humanities. This belief, which I term 'scientism', I felt needed to be explicitly challenged. Another strand is my own fascination with science itself, particularly with my own subject, physics. To some extent, therefore, this book is a sequel to my earlier work *Time, Space and Things*. Part of the nature of science is defined by its origin in the world of Neoplatonic magic of the sixteenth and seventeenth centuries, and in exploring this connection I have benefited from having inflicted on past physics undergraduates at the University of Essex a course of lectures on this topic. Another connection explored is the relation between the residue of magic left over by science and the humanities. My overall aim has been to emphasize that science, for all its power, has definite limitations, and it must yield relevance to the humanities in those areas where its limitations become evident.

In writing this book I have been helped by the suggestions made by Tony Bruce and Simon Critchley and by some excellent copy-editing by Pauline Marsh. Critical readings of first drafts by Sylvia Ridley and by Aaron Ridley have been invaluable for improving clarity of exposition and for avoiding philosophical naivities. What defects remain are in spite of their efforts. They will not, I hope, dilute the conclusion, self-evident to most perhaps, that the cultural world has science and art as complementary aspects and this fundamental complementarity should never be forgotten.

Thorpe-le-Soken, January 2001

One

> The truthful man, in the audacious and ultimate sense presupposed by the faith in science, *thereby affirms another world* than that of life, nature and history; and insofar as he affirms this 'other world', does that not mean that he has to deny its antithesis, this world, our world?
>
> Nietzsche, **On the Genealogy of Morals**

The intellectual activity of mankind is immensely diverse, spanning many dimensions and manifesting itself in count-less disparate forms. A crude but appealing mapping on to one dimension would put science and mathematics at one pole and painting, music and literature at the other, the one commonly characterized by cold, analytical rationality, the other by intuitive feeling and form. From an Apollonian point of view both science and art aim for an understanding of the world; both appear to be part of an all-embracing culture of enquiry, a search for all forms of truth. And off in their own dimensions are religions and philosophies with their own revealed and argued truths. Science, religion, art, philosophy each tend to exhibit the all-too-human myopia of claiming absolute status for its truths, which does not make things easy. Yet, somehow in life, one must make judgements and evaluations – Is this true or false? Is this good or bad? Is this beautiful, or what? There is this urgent need to possess an integration of beliefs about the world.

Looked at in the sense of an all-embracing culture of enquiry, this need would seem to suggest that the methods used so blindingly successfully by science can be applied with equal success to fields as diverse as anthropology and social studies and even art and the humanities. The more fanatical extreme of a belief that this is so is what I propose to call scientism, the religion that given time science will explain all. Science may not take away the sins of the world, but it will certainly describe them truthfully.

One fanaticism brings forth another, if much older one. At the opposite pole is the Dionysian impulse that intellectualizes passionate abandonment in the form of romanticism. The world is the heroic individual, unique, himself; art simply is — a sort of rejoicing, ideally with no other aim; moreover, there is no other world knowable. A contemporary manifestation — post-modernism — is directly antagonistic to the claim that science makes that there is a reality outside the minds of scientists, or indeed in its guise of deconstruction, to any claim of reality underlying any literary text. But these are extremes. Without being ultra-romantic or post-modernist it is possible for humanists to be quite properly dismissive of those extreme claims made for science to be universally applicable in a meaningful, interesting sense. That science may be universally applicable may be simply a truism, but whether it is remotely useful or interesting to apply science universally may be reasonably regarded with scepticism.

Scientism and romanticism are extremes, but they exist as defining elements of a real duality in our intellectual culture, that of science on the one hand and the humanities on the other. Ever since science was distilled from what was a brew of practical crafts fortified with a heady mix of magic and mysticism, and showed itself to be the source of a new and powerful knowledge, it has been perceived by religion and the humanities to be a threat. After Copernicus,

Tycho Brahe, Kepler, Galileo and Newton the Earth was no longer the centre of the universe; after Darwin, man was no longer a direct creation of God. The world appeared more and more to be a world of mathematics rather than of myth, magic and poetry. The threat to the human spirit seemed real. In the 1960s the intellectual world polarized into the infamous Two Cultures of C. P. Snow.

Culture, in this context, was the Culture of Matthew Arnold: 'the best that has been thought and said in the world'. This was taken by many humanists of Snow's generation simply to mean Literature, which, unlike science, celebrates and serves the moral and aesthetic instincts of man. Science may be, and indeed is, a perfectly respectable intellectual discipline, but there was never any need for a cultured person to concern himself with its details. Snow reacted to this by claiming the pre-eminence of science, that science was the only activity capable of advancing knowledge about ourselves and the world we live in. He deplored the narrowness of mind that expected that, of course, everyone has read Shakespeare, but saw no urgency to be acquainted with the Second Law of Thermodynamics and what it implied. There were, Snow claimed, Two Cultures – the literary and the scientific – and there existed a dangerous gulf between them. Snow, scientist and novelist, spanned both cultures and could see the gulf clearly, putting at the door of literature the cause of many of the ills evident in the social and moral life of the nation. This was too much for F. R. Leavis, a literary critic and Cambridge don, who could only be profoundly shocked at science offering its values in place of those of literature. Unhappily, his response to Snow was so vituperative and personally abusive that an opportunity for a reasoned debate on the issue was lost.

Most of the issues raised during the Two Cultures controversy of the 1960s have been thrashed out long ago and may be forgotten, but there are two which continue to be unnecessarily active.[1] One is the virulence of anti-science feeling which informed the intemperate rebuttal that F. R. Leavis made of C. P. Snow's comments on the literary establishment; the other is the perceived incompatibility of literature and science, a perception so deeply ingrained in us by four centuries of antagonism that we have stopped thinking about it and accept it with a shrug. Both appear to stem from the profound psychological effect which the power of science has generated. Because of that power, or at least partly, many perceive that religion has wilted into utilitarian social work, and, for most, battered by the products of information technology, the idea of reverence has become virtually meaningless, and there is the feeling that human existence has lost any conceivable and believable point. In short, science is perceived to be antagonistic to the human spirit. Something is seriously amiss. Surely, nobody would deny that there are scientific works which stand among 'the best that has been thought and written in the world', and yet at least some cultured people appear genuinely afraid of what to them seem to be the consequences of those works, and they tend to reject the excellence. How can a glorious subject like physics and its fellows – chemistry, biology, genetics, etc. – cast such an evil-seeming shadow on the human spirit when they produce some of its acknowledged achievements? How is it that literature, and the humanities generally, which tell us about the richness of human life itself, can be repelled?

The answers to these questions, if, indeed, they exist, will be many and varied, no doubt sad, and not easy to come by. The problem is a complex one involving attitudes to technology as well as to science. How we *apply* science is one thing, science itself quite another. Part of the problem is certainly the confusion of these two activities, the one active in changing our physical environment, the other passive in the sense of expanding the mind's understanding of nature. There are many obvious problems associated with technology which need not be gone into here, real and important though they are – the concern here is with science, and why it often finds itself at odds with the humanities.

Why science should work so well is a mystery in itself worthy of deep reverence – there is certainly no *logical* reason why it should. Its superb power and success have given rise to the tradition that scientific progress in our understanding of the natural world will continue indefinitely, so long as the scientific method survives, a tradition I fully endorse. But tacked on to this tradition, and regarded by many as an integral part of it, is the expectation that this progress will naturally extend to all spheres of human knowledge. This expectation was questioned long ago by Hume, and by others. Hume remarked on the ease with which some people's beliefs slide from what is the case to what ought to be the case, from what is fact to what is value. Writing in 1903 from Trinity College, Cambridge, G. E. Moore noticed that many philosophers conflated natural properties with ethical attributes, that this was like defining the sensation of yellow by its physical equivalent, wavelength. To call something yellow was not to say anything about electromagnetic waves, and to believe that yellow was entirely the same as its wavelength was obviously wrong. Analogous depictions of ethics in terms of natural phenomena were equally fallacious. Moore called this 'the Naturalistic Fallacy'. It seems, nevertheless, the case that the Science Tradition and the Naturalistic Fallacy have taken hold of the mind of modern society with an intensity normally associated with myth, so that any idea of salvation these days is scarcely to be decoupled from science.[2]

It is difficult to understand how the Science Myth can be taken seriously. Its assumption is that the meaningful applicability of the scientific method is limitless. Sociobiologists like E. O. Wilson see salvation in the exploration of the biological roots of morality. Geneticists exploit reductionism to reduce behaviour to that of the genes. Going further, Richard Dawkins sees the evolution of culture itself as the fight for survival of competing 'memes', the selfish genes of culture. Scientists working in the field of artificial intelligence, like Marvin Minsky, see the mind as a glorified computer.

But what can it possibly mean to apply the scientific method to morality or aesthetics, to love and to friendship, to imagination? A whole world of individual everyday experiences lies for ever outside the power of science to investigate in a meaningful way. All the central questions of humanity – Why are we here? What are we for? Is there a meaning to life outside brute existence? – are for ever beyond science, which can only answer questions, if at all, beginning with 'How'. In mathematical terms, to invoke an obvious metaphor derived from the right-angled coordinates introduced by Descartes, the Cartesian coordinates we use in graphs, these questions are orthogonal to science. If belief that there is nothing that science cannot solve, given time, were confined to those ignorant of science, it might be bearable, but very unhappily that is not the case. It is sometimes intellectually embarrassing to observe eminent fellow scientists abandon the imaginative and subtle thought which they display in their own fields and cheerfully promulgate this appalling myth. The worst are often reductionist biologists who are either able to gloss over enormous semantic cracks and see morality in a gene, or who regard morality as meaningless. But there is not much to choose between them and enthusiastic members of the artificial intelligence (AI) community, who believe that people are basically computers. Physicists, I am happy to say, are somewhat cannier, by and large, perhaps because of the humbling effect of the discovery of the deeply non-intuitive phenomena of the quantum world, but in cosmology there have been unhappy examples – a claim of the possibility of a Grand Unified Theory of the Universe slipping off-handedly and notoriously into an unqualified claim of the possibility of knowing the mind of God (that is, if an omnipotent, absolutely perfect being has such a mudane thing as a mind). If the no-bounds myth of science is entrenched in the attitudes of our brightest and best, it is scarcely surprising to find it in society generally. Some of the rot is in science itself, and justified more often than not by the dismissal of all human values, all questions beginning with 'Why', as meaningless, merely because they lie outside science. This in spite of the fact that concern about value and 'Why' is empirically observable and is as much part of the world as a molecule is, and infinitely more richly experienced. That these concerns lie beyond the power of science and that science has real limitations seem hard to bear for a certain cast of mind.

If a trigger were needed for airing the problems raised by scientism, one can be found in the plethora of books containing claims of one sort or another about the power of science – either it is all-potent, or it is ending.[3] The claim that it is ending cannot be countenanced seriously (though bits of physics may be suffering from hardened arteries), given our enormous ignorance about the world. If the submicroscopic nature of matter and the broad structure of the universe look from a certain standpoint fairly well sorted out, the world of the gene is not. The selfishness of the gene is a wonderful metaphor, whatever its usefulness for bridging the vast gulf between based genetic properties and animal (including human) behaviour.[4] Equally, the all-potent claim is certainly extravagant, even if science is sometimes described as the most powerful intellectual activity of the human race. But the claim that, given enough time and resources, science will deliver on anything, and the even more depressing claim that whatever cannot be scientifically described is meaningless, needs to be continually challenged. One is reminded of the erudition of a certain Master of Balliol:

> First come I; my name is Jowett.
> There's no knowledge but I know it.
> I am Master of this college:
> What I don't know isn't knowledge.

Claims of the mock Jowett sort give science a bad name. There are too many obvious limitations to the scientific method.

A brief look at the most important of these limitations is in order at this point. Science is, above all, a collaborative venture which is world-wide. It is utterly alone among the intellectual disciplines in that respect. Its fundamental feature is the unambiguous communicability of its findings, and herein lies its undoubted strength but also its fundamental limitation, in that it can deal only with knowledge which is so objective, so testable, so repeatable and so specially public that it is meaningful to anyone engaged in the scientific method, wherever he may be. Natural phenomena possess all sorts of unique idiosyncratic features in addition to features which are scientifically accessible. Anyone who has ever carried out an experiment will know what I mean. Reducing the effects of those unwanted elements – extraneous electromagnetic fields, traffic vibrations, power surges, union action and, in field-work, even beetles in the galvonometer, etc., etc. – tests the skill of the experimenter. Only what is repeatable is extracted from the rich activity of nature and converted into publishable science. Scientific knowledge is a special abstraction of what is presented by nature. It cannot have anything whatsoever to do with the unique, the unrepeatable. Yet nature is rich in unique, unrepeatable happenings that it transcends language to describe.

It has to be said that scientific knowledge obtainable by the intellectually pure, ascetic and certainly, in a sense, chaste, activity of science is the *simplest* available. In the empirical sciences, what is known confidently is what is repeatable. A phenomenon observed in California has to be observable in Japan, or if it is only observable in one place, like an eclipse, there must be many instances. Scientific knowledge is not only *public* knowledge[5] – most knowledge, after all, is public knowledge – it is the simplest. Unique events in the universe, which nevertheless exist, are for ever outside its ken. It cannot do otherwise than focus on a tiny element of reality, that is, to analyse into manageable interactive components what is sometimes clearly an organic whole. Holistic features cannot be apprehended. The more instances, the better the physics can be. But think of the number of unrepeatable events. *Every event is unrepeatable!* Scientific knowledge is sound in proportion to its ability to discount the effect of those elements of events which are unrepeatable. But by doing so it throws out babies galore with the bath water. Science retains its air of chaste virginity only by the fastidious avoidance of the importunate behaviour of the real, rapacious world.

In short, it is as if there were a pact which men and

women of empirical science make with the physical world. We promise, they say, never to use irrational abilities such as intuition and imagination without discounting the fact with a comprehensive display of rationality. Furthermore, we accept that the consequence of that approach will be to limit what can be believed as true to those special aspects of the world, the ones which recur. In return the physical world promises that *there will be no event without an element of recurrence*. It almost follows that the simpler the event, the larger the element of recurrence, for that is the case. All electrons look alike – good for elementary-particle physics. All atoms of tungsten look pretty much alike – good for atomic physics. Large molecules and crystals begin to be a little idiosyncratic, an atom misplaced here and there, but still they are largely predictable – moderately good for solid-state physics and chemistry. Huge organic molecules in living cells, cells in microscopic organisms – getting difficult for biology. Organizations of living cells in animals, organizations of animals in societies – worse and worse. Consciousness – well, there are still elements of recurrence even here, and hence psychology, but interesting truth is very difficult to come by at this level of complexity.

Besides the testable truth of science, the demonstrable truth of mathematics, the revealed truths of religion and the persuasive truths of the humanities, there exists, I contend, another kind of truth which is – what better way to put it? – a magical truth. The non-material human forces in this world are magical forces in the sense that they are non-mechanical and indescribable by science. They are the power in personality, charisma, ritual, form, atmosphere – the effect of the animate and inanimate on the mind. The power of magic is what art is about. It is exploited in religion, in rhetoric, in patriotism. It is there in the home and in the garden. This is magic that should not conjure up disreputable images. What I mean by magic in this context is not conjuring-trick magic or black magic, or superstition, but what, long ago, was called natural magic, and this has its truths outside scientific truth. This power, these truths are known intuitively or not at all; they belong to the purely human world of sensibility. Magical truth and scientific truth are complementary in their respective limits, the former associated timelessly with the unique, the latter timelessly with the recurrent. The timelessness is all. At one time to be educated was to be well versed in both, but one approach does tend to drive out the other. The scientist as such can never be a magus (though as a human being he may be!). Poetry, music and the fine arts embody magical truths. That is where to find them, abstracted. And magic, I believe, is the right word to use. At one time the activities of science and magic were virtually indistinguishable. The magic I mean is the potent liquor left behind by the distillation of science, freed from superstition and as near an elixir of life as we are likely to get.

What is not scientific is not necessarily superstition. Forces that move people exist which lie outside the scientific domain by their very nature – everyday forces, neither supernatural nor occult, plain to everybody, part of the human experience. Even our everyday language employs the image of the mechanical effect of a force when we speak of being moved by a certain purely mental experience. Who does not respond to the power of form, of colour, of symbols, and is this not what used to be called talismanic magic? Who is not delighted and moved by the artful use of words, incantations, names as essences, oratory, poetry, and is this not word magic? Is not the effect of harmony and melody magic? And is there not an elemental magic in the intuition of the craftsman, even in the 'feel' of the technician for his machine, though it be a product of the highest technology? To say nothing of common sense? Surely the answer to all these questions is yes. Then these are nothing but the well-known elements of natural magic stripped of their superstitious and supernatural patina. Magic, thus defined, is the complement of science. It acts, not on material objects, but on human sensibility. At its highest intellectual development, reached in the seventeenth century just when science was beginning, magical theory could describe a cosmos full of meaning to the human spirit, a latter-day Theory of Everything.

One of the earliest discoveries of the magical age was that the world was Number and Proportion. Ever since the Pythagoreans elevated mere counting into a philosophy, mathematics has played a fundamental part in cosmology, from the days when the cosmos and music were one down to the present, where now the music of the spheres, if discernible, has a distinctly general-relativistic tensorial structure. The connection between magic and mathematics and the connection between mathematics and science are too close to ignore. That mathematics can be used to describe the physical world is, nevertheless, a deep mystery.

An accurate, passionless description of the physical world entails its own morality – the truth is discoverable only by the truthful. But what of the world we immediately know about, not the abstract world created by science, but the world in which we, including scientists, actually live? There is no simple morality here, but there are values: material, aesthetic, ethical. They exist because consciousness exists. They reflect the nature and needs of the mind and body and of society. At any time, a unique set of values motivates and moderates the actions and judgements of the individual, and form a vital part of his faith. Some of the material values, those to do with self-preservation and basic needs, are common among all but the insane and sick, and are as absolute a set as one can get. And, no doubt, there exists a deeply buried connection between the properties of the human genome and elements such as ethics and language, and, indeed, between a whole host of so-called epigenetic attributes. That apart, values, like faith, are creatively evolvable. Wisdom is to know not only 'thyself' but also your neighbour, your boss, your local party chief, your physical, economic and political situation, and everything besides. Wisdom is to create a faith-value package accordingly, and then to recognize wryly how close that faith-value package is to the one which, as it turns out, you happen to have.

Science and faith, magic and engineering; knowing, believing, manipulating, acting; such is life. The world seems irrevocably divided into a world of fact and a world of value. On the one hand there is matter subject to the laws discovered by physics; on the other, there is – for want of a better word – spirit. (The trouble with the word spirit is that it has an entrenched aura of the supernatural which I emphatically do not want to evoke. There is nothing supernatural about conscious life, is there?) They represent two entirely different categories of being, between which there can be no easy discursive account. Not that that has stopped people trying. In the fifteenth and sixteenth centuries matter was commonly spiritualized, and indeed the whole practice of Hermetic magic depended on this idea. In the present century it is common to find spirit materialized, with desperate attempts being made to explain mind in terms of a quantum mechanical brain.

Yet matter and spirit are part of the same world and

our understanding must attempt to encompass both. The old dualist, idealist and materialist solutions are no longer interesting, though some sort of Spinozan dual-aspect theory may yet be. Spinoza saw nature as One. His monism came directly from his conception of substance as having attributes that are intrinsic and independent of anything else. It logically follows from that definition of substance that there can only be one such substance – God or nature. Mind and matter then must be aspects of the same substance, as heads and tails are the double aspects of a coin.

Refining the Scientific Myth is a challenge that has been taken up by science in recent years but, inevitably, the understanding sought is materialistic in nature, even if the materialism is much more sophisticated than it used to be. At one pole is a renewed attempt to understand the quantum nature of the world. At the other, and most ambitious of all, is the attempt to account for man's place in the universe in terms of Anthropic Principles and to explain the origin of the universe itself. Man himself is emerging as a biochemical entity whose abilities are circumscribed by his genetic make-up and whose very moods are chemically controlled and whose brain is a glorified computer. The search for a Theory of Everything is in the air, which in physics is dangerously close to converting the subject into a kind of mathematical theology almost entirely divorced from an empirical base too expensive to maintain.

That movement in science that addresses topics traditionally treated by religion bears directly on the faith-value package that each of us has. Because of that we need to evaluate rather carefully what this new materialism is telling us. Physics is fundamental, and we need to understand the new thinking about the quantum world and the trend towards mathematical theology. We need to ask what the limits of science itself are and how far its divorce from magic has limited its ability to explore the conscious universe. And, finally, we need to ask to what extent this new thinking offers a substitute for religion and what it has to say, if anything, about the existence of God. This book addresses these concerns, but its principal aim is to restate the blindingly obvious point that human sensibility is responsive to forces that lie outside the regimes of physics and biochemistry and whose description cannot be reduced to those regimes. It would follow that any cultural integration of the arts and the sciences must take that fact on board, and if this means accepting an intellectual double aspect in human culture, then so be it.

The Limits of Science

Two

The enlargement of insight in mathematics and the possibility of new
inventions extends to infinity; equally the discovery of new properties of
nature, new forces and laws, by continued experience and unification of
it by reason. But none the less we must not fail to see limits here, for
mathematics only bears on appearances, and what cannot be an object
of sensible intuition, such as the concepts of metaphysics and morals,
lies quite outside its sphere.

Kant, **Prologomena to any Future Metaphysics that will be
Able to Present Itself as a Science**

Four hundred years ago in Western Europe the business of
understanding and controlling nature lay mainly in the hands
of magicians – the astrologers, alchemists, Hermetic philosophers, Rosicrucians and the like – an upper-class example
being Shakespeare's Prospero. If there was a secular theory of
the world, it was founded on Magic with its spirits and
demons, occult sympathies between one bit of the world and
another, emanations descending from the celestial sphere,
man as microcosm and so on. Out of this miasma, science
slowly crystallized and developed and became the enormously powerful and successful activity with which we are all
familiar. Magic and all grosser irrationalities were gradually
abandoned, and religion had to examine its beliefs and reach
a not altogether painless rapprochement with the new rationality. Art largely ignored the whole affair, for reasons that
were quite respectable – what, after all, had the discoveries of
science to do with art? Technology has changed our material

On Science

lives out of all recognition, and the idea of science, its quantitative methods, its perceived logical and analytical approach to the world, now permeates society and its institutions with an intensity inconceivable even a few decades ago.

In the seventeenth century, Giordano Bruno, Johannes Kepler and Galileo could and did shock the Roman Catholic establishment with their ideas. The so-called Enlightenment of the eighteenth century saw the attack on religion by people like Hume and Voltaire, but at the same time heard the devout masses of Bach and saw prominent scientists like Joseph Priestley, the discoverer of oxygen, devote themselves to transcendent visions. In the nineteenth century there were devout Christians studying electromagnetism, like James Clerk Maxwell and Michael Faraday, and the publication of Darwin's book *On the Origin of Species* caused a religious furore difficult to imagine today. Even in the years during and following the Second World War there was no shortage of transcendent fervour among practising scientists. The code-name of the atomic bomb project was Trinity, the father of rocketry, von Braun, was intensely religious, and Einstein had a vision of a God who certainly did not play dice. Nowadays, churches in the West, the ones that still exist, are generally much emptier than they were. Religious fervour has been largely metamorphosed into deep feelings about the environment or about animal rights or whatever. Issues like the rights of abortion, euthanasia and eugenics still evoke responses from institutional religion, but practical problems in these areas tend to be resolved more by down-to-earth ethics, or by individual action, than by any appeal to God. If scientists these days evoke God, it is almost certainly an evocation arising from a feeling of awe generated by the study of nature rather than one generated by the study of the Bible or of the Koran.

Western society has never been more secular and materialistic, and it seems to more and more people that ordinary human values informing the aesthetic, moral and spiritual life of the individual are counting for less and less. There have been, and still are, scientists and philosophers who define everything that is beyond the scientific method as meaningless, and such utterances hasten the process of (literally) devaluation. A whole school of philosophy was created by the Vienna Circle, which formed around Rudolph Carnap in the years between the wars. Its concentration on language and its meaning is held by many to have blighted philosophy for years. Its central idea was to define the meaning of a scientific statement as the procedure for verifying its truth. If an experiment could not be carried out to test a claim, that claim was meaningless. A few years ago Rolf Landauer, a highly talented physicist in IBM's research laboratory, told me that he believed that whatever could not be measured was meaningless, so the essence of verificationism is still alive and well. Such views are bad for science. The evident powerfulness of science, perhaps inevitably, has created an anti-science strain in society that associates a perceived spiritual impoverishment with the huge success of science and, not stopping there, blames science for various physical ills, often associated with its technological application.

Yet science is clearly limited in its scope by its own methods. An obvious limitation is its need to disturb in order to measure. The idea is to keep that disturbance as small as possible, but sometimes that is not possible. Many examples exist in the quantum regime, but it can also happen in measurements involving people where the act of measurement alters the behaviour of people. It is a limitation frequently remarked upon and just as frequently forgotten. Without an awareness of this it is impossible to obtain a balanced assessment of the rôle science does and can play in our culture. Such is the present-day aura of the subject that I dare say that it may come as a surprise to some that bounds actually exist. Reading some articles that have appeared in the press in recent years – quality press as well as tabloid – for example, about the risks of humans catching 'mad cow disease', which are small compared to catching almost any other horror, one might be forgiven for thinking that there are a number of (presumably) educated writers for whom science and magic are still intermingled. Even among bright undergraduates one can come across the belief that, given time, science will solve everything. On the contrary, it is obviously and inescapably limited. Though the methods it uses are the best that can be imagined for the purpose of gaining an understanding of the 'natural world', through its use of reason, insight and imagination, nevertheless any view of science as purely rational, unbounded in scope and limitlessly powerful is a myth which for science's sake and for society's should be scotched (and often has been, but with little cultural awareness).

The view of science as an activity that generates 100 per cent absolute truth was, of course, discredited long ago by Hume, when he pointed out that knowledge gained through experiment could never be certain.[1] Even though nature had behaved hitherto in the way described by a theory drawn inductively from observation, there was no logical compulsion for it to continue to do so. Our belief that it will is exactly that – a belief. All our science is based on that belief – reason does not come into it. Kant's attempt to counter Hume's criticism by postulating *a priori* insight, such as of Euclidean space, could not survive subsequent discoveries, such as non-Euclidean geometry.[2] (But leave out 'Euclidean' and Kant has a timeless point – try thinking of a lack of matter without thinking of space.) Nevertheless, it was Kant who emphasized that empiricism had to do with the appearance of things only and not with the things-in-themselves; these must remain for ever beyond our ken. He was at pains in his Prolegomena to point out that confusing appearances with things-in-themselves leads to nonsense. Thus the claim that the world, as to time and space, has a beginning and the counter-claim that the world, as to time and space, is infinite are both meaningless, since infinite space or time, or bounding the world, are ideas about things that are outside any possible experience. So much for some modern cosmology!

There is an apparently unbridgeable gap between our theories and what things are really like. The point here is the paradoxical one that science ultimately does not know what it is talking about. As scientists we can grasp ways in which nature behaves, but what true reality is like remains a mystery. Many would argue that our theories are merely instruments for manipulating the world and that they embody no truly fundamental insights. Some would adopt the more extreme view, more conducive to literati and social scientists of a post-modernist persuasion than to physical and biological scientists, that science is nothing more than the activity of publishing papers according to some convention or paradigm whose claim to describe a real world is nonsense. In any sane view, the overwhelming success of technology – actually manipulating the world – gives these scepticisms the lie. We cannot prove that there exists a real world which corresponds to our theories, but it is impossible in our bone-marrow to doubt its existence.

Nevertheless, do atoms really exist? Do electrons really

exist? Most of us believe so – we are all scientific realists in the laboratory – but it is a belief originating more from pragmatism than Kantian insight, it has to be said. As physicists we are as familiar as we can be with forces like gravitation, electricity and magnetism, but however familiar, these forces are ultimately mysterious. They demonstrate action at a distance. We have learnt how to manipulate them, but what are they really? Why should mass attract mass? What is electric charge? Is it meaningless, as positivists claim, to wonder what the world is really like? Frankly, childlike, I would like to know what gravity really is, but I know that science can never tell me. It can tell me how, yes, but never what or why. But, in this, science is in no way different from anything else, unless it be religion.

So what is science? I cannot do better than quote from Edward O. Wilson's book *Consilience*.[3]

Science, to put its warrant as concisely as possible, is the *organized, systematic enterprise that gathers knowledge about the world and condenses the knowledge into testable laws and principles.* The diagnostic features of science that distinguish it from pseudo-science are first, repeatability: The same phenomenon is sought again, preferably by independent investigation, and the interpretation given to it is confirmed or discarded by means of novel analysis and experimentation. Second, economy: Scientists attempt to abstract the information into the form that is both simplest and most pleasing – the combination called elegance – while yielding the largest amount of information with the least amount of effort. Third, mensuration: If something can be properly measured, using universally accepted scales, generalizations about it are rendered unambiguous. Fourth, heuristics: The

best science stimulates further discovery, often in
unpredictable new directions; and the new knowledge
provides an additional test of the original principles that led to
the discovery. Fifth and finally, consilience: The explanations
of different phenomena most likely to survive are those that
can be connected and proved consistent with one another.

Here are all the main features clearly set out: repeatability,
economy, quantification, stimulation, internal coherence. It
is noteworthy that aesthetics enters via the concepts of
theoretical elegance and economy, a pointer to the existence
of a metascience. There is, however, one important missing
feature, but this is soon added:

The cutting edge of science is reductionism, the breaking
apart of nature into its natural constituents . . .

Limitations are imposed by two aspects of scientific practice, namely, analysis (reductionism) and the inevitable
necessity of using one bit of the physical world to measure or
investigate another bit, which is a kind of self-referentiality.
The problem with analysis, without which no progress can be
conceived, is the danger of missing the whole that is greater
than the sum of its parts. As Schrödinger put it, you cannot
reduce a man into individual elementary particles without
killing him; something might therefore be lost in the description.[4] The same criticism might be levelled at all attempts to
reduce the description of man to that of his molecules or
genes. Nor need such a criticism of reductionism imply anything supernatural in man. Analysis proceeds at many levels. It
is inevitable that the language of one level is likely to be
inappropriate at another. Solid-state physicists do not talk
about quarks when they describe crystal structure. Software

engineers do not talk about transistors. It is therefore striking that molecular biologists have a tendency to extrapolate their findings many levels up to the world of the spirit.

Reductionism is endemic. The scientific world is made of atoms, atoms are made of electrons and nucleons, nucleons are made of quarks; heaven knows what electrons and quarks are made of, but it would be an extraordinary physicist who contemplated a seamless linkage of these ultimate objects of reductionism to the phenomenon of human consciousness. Yet the idea of the world as a unity, that persistent intuition, the characteristic of long-discarded magical theory, demands that such a linkage must exist, and it is an idea that motivates the thinking of many scientists in the fields of genetics and computer science. Wilson's book expresses this sentiment from his standpoint of sociobiology:

> The central idea of the consilience world view is that all tangible phenomena, from the birth of stars to the workings of social institutions, are based on material processes that are ultimately reducible, however long and tortuous the sequence, to the laws of physics.

This is a clear statement of the reductionist manifesto.

Among the first steps in this sequence would be atoms to molecules to large molecular structures with the ability to replicate in some primeval soup. There is little doubt that some such mechanism must be responsible for the origin of life, and one day interesting replication may indeed be demonstrated. Beyond this, things get very complicated. A successful replicator will have to develop some sort of protective skin or armour against other successful replicators which at the same time does not seriously impair its powers to reproduce itself. There will have to be trade-offs that will still allow it to develop adaptability to environmental changes to the food supply. And so on and so on to living cells with DNA, genes and chromosomes. Linking basic replication of molecules to the gene is a tremendous challenge to science, and it is challenges of this sort that make science exciting. Outside fundamentalist religion, there are few who doubt that such a programme is feasible and worth while and will be ultimately successful.

That certainty is the belief of those biologists, perhaps all biologists, whose reductionism of life begins with the gene. They can safely leave the links involving the replicating molecule to chemistry and they can concentrate on how genes affect behaviour, if at all. Richard Dawkins argues for the gene inheriting that fundamental replicating power, summarized metaphorically by the term the selfish gene.[5] For this idea to work the behaviour of animals and humans must somehow be determined in such a way as to optimize the chances of survival of the gene. Since there are numerous genes, each with its specialized rôle within the genome and its own optimizing strategy for survival (to pursue the metaphor), it is not obvious how any conceivable human behaviour could ever be unambiguously tied down to the survival action of one gene. This has to be especially so when the time-scale for evolutionary survival is taken into account. Even if survival is in the context of changes in social culture rather than changes in climate on a geological time-scale, the period involved will be at least several generations long, and this will render the possibility of a scientific conclusion being drawn, independent of rhetoric, difficult, to say the least. Nevertheless, it is true that genes exist and that they determine elements of a person's make-up, colour of eye, facial features, and so on. But one only has to recall the virulence of the nature–nurture debate on an attribute like intelligence to see how difficult it is to substantiate any claim for genetic causation of behaviour, even behaviour as widespread in different societies as the avoidence of incest.

Nevertheless, it is in the nature of scientists to be epistemologically optimistic, and nowhere is this more clearly evidenced than in Wilson's book *Consilience*, in which the author sees that:

> the humanities, ranging from philosophy and history to moral reasoning, comparative religion, and interpretation of the arts, will draw closer to the sciences and partly fuse with them.

And further:

> There is only one way to unite the great branches of learning and end the culture wars. It is to view the boundary between the scientific and literary cultures not as a territorial line but as a broad and mostly unexplored terrain awaiting cooperative entry from both sides.

But he is realistic enough to anticipate one response:

> [philosophers] will draw this indictment: conflation, simplism, ontological reductionism, scientism, and other sins made official by the hissing suffix. To which I plead guilty, guilty, guilty.

He is right about reductionism – science cannot be science without it – and he finds stimulation in Dawkins' idea of the cultural gene,[6] the meme, examples of which are:

> tunes, ideas, catch-phrases, clothes fashions, ways of making pots or of building arches.

Memes are the new replicators, perhaps elements of actual neural structures in the brain that replicate in one brain after another. The meme–gene connection, as far as I can see, appears to be encapsulated by the syllogism:

Genes prescribe epigenetic rules.

Culture helps to determine gene survival.

Therefore, successful new genes alter epigenetic rules and change the direction of culture.

The claims of sociobiologists like Wilson are vast. Extrapolating from the study of insect communities, they see evolutionary sociobiological principles as explaining the whole human world of morality, altruism and all the other cultural forces. Kant's categorical imperative, Nietzsche's analysis in his *Genealogy of Morals*, G. E. Moore's intuitive perceptions of morality, are all seen to be missing the point, the point being that morality and all cultural forces develop and evolve in order for the genome to survive. Socio-environmental events, like people talking to each other, writing about one another, governing one another, do not determine culture. Culture is genetic.

For me, these claims are simply not credible. They are a product of unbridled scientism. Such views do not make for easy, or any, conciliation with the humanities. But with those claims toned down and sociobiological analogies more modestly advanced, the programme of conciliation on the science side looks lively and interesting, but it is difficult to see it matched on the other side by anything that Wilson would recognize as research.

The problem, as I see it, is that the two cultures are complementary in the same way as, in quantum physics, momentum and position, energy and time, are complementary. Focusing on one drives out the other. Our earlier metaphor, drawn from coordinate geometry, is that the two cultures are mutually orthogonal; the idea of a territorial frontier between them assumes that they are more alike than they really are. It is true that, as far as scholarship is concerned, the same attributes of careful research, truthfulness and public exposure of work done exist on both sides. But there are also big differences which are not recognized by, for example, the definition of art that Wilson gives:

> Art is the means by which people of similar cognition reach out to others in order to transmit information

or:

> The common property of science and art is the transmission of information and in one sense the respective modes of transmission in science and art can be made logically equivalent.

All of that puts art in a utilitarian, measurable category, in which it most definitely does not belong. Think, for example, of Beethoven's Ninth, Botticelli's *Prima Vera*, Durham Cathedral. What art is is a question that has bothered philosophers from Plato to the present day, but, as R. G. Collingwood argues convincingly,[7] art is not a craft. Craft has an aim and knows when it has achieved that aim. In Collingwood's view, an art that has the aim of transmitting information would be misnamed − it would be a craft. Of course, reducing art to craft makes the connection with science more plausible. But if art has any aim at all, it is a kind of celebration of the human condition by the artist, whether it is shared by anybody else or not. We will return to the relationship between art and science later, but it is sufficient at this point to discount any attempt to reduce art, or at any rate, all kinds of art, to a kind of information technology.

A note in passing: reductionism in art is not unknown. Mozart, for fun, wrote music that was to be constructed by throwing dice. The idea is that one of a set of introductory bars chosen by the throw of a die is coupled to one of a set of answering bars, again chosen by throwing a die. The result is a finished little piece. We have here, possibly, Mozartian memes of music.

One of the paradoxes that confront reductionism in the microscopic world is that the process of analysing the world involves the separation of the bit to be studied from the rest, and quantum theory as it stands at present informs us that this is impossible. The discovery that the behaviour of subatomic particles like the electron and proton was very different from that of small billiard-balls led to the development of quantum mechanics in the 1920s and 1930s. After a heady mix of confusion, controversy and brilliant mathematics associated with the names Niels Bohr, Werner Heisenberg, Albert Einstein, Erwin Schrödinger, Max Born, Paul Dirac and many others, a view of what this new physics was all about eventually crystallized. Heavily influenced by Bohr's positivist approach, this view became known as the Copenhagen Interpretation and was broadly accepted by working physicists. In it all thought of describing subatomic reality was abandoned. The equations of the theory were merely instruments for describing the result of quantum particles interacting with macroscopic bits of equipment. The measurement became the message.

The physical world became more and more mysterious. Quantum systems turned out to be essentially holistic, being described by a single wavefunction which describes its dynamic state. The interaction with some measuring system means merely that the quantum system is bigger than we at first thought, now including the measuring device. The wavefunctions of the system to be measured and the system doing the measurement get entangled. Moreover, the measuring system is in touch with the rest of the universe, and so the totality is described by a Grand Universal Quantum Theoretic Wavefunction. According to Schrödinger's equation all dynamic possibilities continue to evolve in a deterministic manner, yet when we carry out a measurement only one of these possibilities is realized. Saving the phenomena for many physicists means believing that the universe continually divides to contain all possible results of measurement. The Many Worlds (or in a related view, the Many Minds) interpretation grants a reality to the wavefunction and an epistemological primacy to Schrödinger's equation that is as absolute as it is remarkable. Such an interpretation of quantum theory has an air about it that is both bizarre and desperate. A quantum system certainly possesses this curious feature of non-local entanglement, and its wavefunction certainly describes the possible outcome of a measurement of a particular physical quantity like position or momentum, but once it interacts with a macroscopic measuring device we get a definite result; only one of the possibilities is realized. We refer to this as the *collapse of the wavefunction*. Without this collapse of the wavefunction, analysis of quantum systems, or indeed of anything, would be impossible.

The Copenhagen Interpretation regarded all of this with equanimity. What the wavefunction described was only a kind of betting odds to be understood in the context of a particular measurement. The underlying reality was,

Kant-like, for ever mysterious. Any measurement was bound to use macroscopic equipment that could provide only results definable in the terms of classical physics. So there was bound to be some disjunction between the microscopic and the macroscopic world and hence the apparent collapse of the wavefunction was merely a symptom of this. In the everyday application of quantum theory nobody worries about all this, but to many it is an unsatisfactory feature of the theory. What causes the wavefunction to collapse? There is nothing in the standard theory to answer this. Quantum mechanics, the fundamental theory of the physical and therefore the scientific world, appears to be incomplete. It cannot say how far analysis (in the present context – measurement) destroys the whole. The unperturbed quantum system is forever mysterious. In fact, the basic quantum nature of the world is still very poorly understood.

The other aspect, self-referentiality, closes science in upon itself. The things of nature are used by science to investigate the things of nature. The famous example is the Theory of Relativity. The Kantian Idea of space and time gets an operational definition. In order to map out time and space so that dynamic events can be measured, we use another dynamic event – light – and we choose light because it is the fastest carrier of information that we know and it has the virtue of penetrating a vacuum. Our clock is a vibrating caesium atom and we define a relativistic world by defining a value for the velocity of light and defining any path that light takes to be the shortest. The world so described is a world of curved space-time quantitatively defined by the method used to obtain the information. It is not absolute. Were we to discover an information carrier faster than light, it would reveal a different world. It would still be a world revealed by itself, described in terms of itself, and ultimately never going outside itself.

Science is essentially a description of the motion of matter. In physics the describing is done in terms of differential equations, which, of course, will yield solutions only if the boundary conditions are known. There are a great many cases where boundary conditions cannot be defined precisely. (In certain non-linear, dynamically unstable systems no amount of precision will allow of an accurate description of motion beyond a certain time.) In predicting the dynamic state of a gas we cannot possibly know the starting position and velocity of each molecule. We therefore impose *a priori* ideas of probability on to a model, and thereby introduce the concepts of chance and randomness. Chance rules out purpose, and choosing chance is a deliberate choice which serves science very well in that it works. The discovery that the concept of probability was not merely of use as a cloak for our ignorance as in classical physics, but was essential at the fundamental level of quantum particles, meant that self-consistent statements could only ever be made about the average behaviour of large numbers. There are a few billiard-ball-like systems in nature where the statistical element is insignificant, but there exist many systems where only a statistical description is possible. Statistical ideas are meaningful only if applied to large ensembles of identical systems, such as may be said to exist in the case of a gas of atoms, but the concept becomes increasingly irrelevant the more complex and rare the system becomes. More and more the idea of a large ensemble of identical things becomes a vision rather than an approximation to what most of us think of as reality, when the element is a complex one such as an animal, or a man, or a society, or a whole planet. Stephen Hawking once pointed out that any

Grand Unified Theory of the Universe that takes into account quantum theory must be statistical in nature, implying not one universe but 'an ensemble of all possible universes with some probability distribution'.[8]

Outside plain mathematical instrumentalism it is difficult to understand what epistemological significance statements of this sort have, for three reasons. One is the problem attending any idea of completeness in science such as that implied by a Grand Unified Theory or a Theory of Everything. Insofar as such a theory is mathematical and can be quantified arithmetically, Gödel's Incompleteness Theorem applies.[9] Kurt Gödel's paper, published in 1931, shocked the mathematics community to its core. It was felt by many inside and outside the subject that mathematics could be formalized in terms of pure logic, or at least as a system based on a finite number of premises from which all mathematical truths could be deduced. But Gödel showed that this intuition was simply not true: any mathematical system as well formed as arithmetic could have truths not derivable from any finite set of premises.

If Gödel's theorem does apply, there may be certain truths of nature, quantitatively expressible in arithmetic terms, not predictable by any theory with a finite number of axioms but nevertheless discoverable. This in itself does not vitiate the search for a Theory of Everything, merely means that it may turn out to be only a Theory of Nearly Everything.

A second limitation, in a similar technical sense, is complexity. More and more problems are tackled these days by computer modelling, but is that always going to be possible? If a quantitative description of some complex physical system is wanted, a mathematical model is prepared in the form of a set of instructions for the most powerful (fast, large-memory) computer available. One feeds the program in and sits back waiting for the computer hardware to do its job and finally produce a result. The question is, can this always be successful? Can it be proven that, given any arbitrary set of instructions, the computer will always produce a result? The bother is that it is known that some simple programs can be written that would never allow a computer to halt eventually. Alan Turing, a few years after Gödel published his bombshell paper in 1931, proved that it is impossible to tell in advance whether any set of arbitrary procedures fed into an infinitely powerful computer – a Universal Turing Machine – will lead to the machine eventually stopping and delivering a result. This is known as the Halting Problem.[10]

The third reason is to do with the question of uniqueness. We now reach the most serious of all the limitations of science – its inability to cope with anything unique. The limitation is self-imposed – scientific knowledge must be globally public knowledge. It must remain valid, or at least testable, throughout the laboratories of the world. As such it must deal with events which are describable and repeatable. Now, every scientific experiment is a unique event, carried out at this precise time, at this location, with this particular piece of equipment, by that genetically unique experimentalist. Every scientific experiment is unique and unrepeatable. The science that the experimentalist can publish is what can be extracted that is repeatable. All unique and subjective elements must be purged from the statements of what has been achieved. In this way science abstracts from the enormously complex set of events that make up nature just those elements with a sufficient degree of repeatability about them to make them candidates for firm scientific knowledge. It is a highly productive and successful strategy that has led to an impressive understanding of just those parts of nature which fit science's criteria. One may ask if science cannot cope with unique things, what is it doing attempting a description of the universe? Whatever mathematical conceits there may be concerning an ensemble of universes, the fact is that our experience is of only one universe, by definition. Nothing, in other words, is more unique. Yet cosmology exists and is truly fascinating. But is it science, or what? It is certainly science in its treatment of observables like the relative abundance of the elements or the cosmic microwave background, but its speculations concerning genesis and eschatology are less obviously so, however fascinating.

Nature is rich in exactly those elements where purely mechanical motion predominates. Its richness in this respect falls off noticeably in biology and all but vanishes in man. Science, by its nature, can have nothing interesting to say about individual human values. Aesthetic and moral experiences are highly complex and highly subjective, with aspects that are essentially unique to the individual. They are, nevertheless, experiences and part of nature, but such unique experiences are for ever outside science's ken. Art works are also unique. Not that it would seriously dream of doing so, but what can science say of interest about Beethoven's Ninth, the Prima Vera, or Durham Cathedral?

But in many fields science has a counter to the criticism that it cannot cope with the unique. All right, it says, let us accept that unique events exist. There are always going to be odd, unpredictable patterns of interaction in nature because nature is so complex. But these rare or even unique occurrences do not matter in the long run, because whatever uniqueness flows from them is soon destroyed or overwhelmed by the tendency of nature to favour the norm, the average, the general. Nothing in the world is irrevocably changed by uniqueness, and therefore unique events are ignorable.

A more potent counter argument is that there can be no objective account of any kind of truly unique things, so science is not alone in having this limitation. Public, communicable elements must exist for concepts and theories to exist, and that applies to art, morality and any human activity. The difficulty of developing objective accounts of aesthetics or morality, for example, is dealing with the rich variety of human responses that are possible, and science cannot, in principle, deal with that. Aesthetics can aspire to objectivity because, for example, there exists music other than Beethoven's Ninth, paintings other than Botticelli's Prima Vera, and cathedrals other than Durham. Science is limited here because of the large presence of uniqueness, in the elements of human culture. But if uniqueness cannot, by its very nature, be treated objectively, it is still a vital component of the world. It cannot be ignored. Every human being is unique, and so are his responses to unique things, even if talking about them entails the existence, or the development, of a suitable public language. The language of science is certainly public, but it is scarcely suitable for describing the aesthetic or moral experience.

The fact of the matter is that scientific knowledge is not the only knowledge, nor is it necessarily always the most interesting knowledge. It is a knowledge only of accurately repeatable things that can be described unambiguously using clear, nonmetaphorical language. It is, above all, knowledge by description. But, as Russell reminds us and as we all know by direct experience, there is also knowledge by acquaintance. Acquiring the ability to distinguish a Bordeaux from a Burgundy, or at a more elevated level, developing an appreciation of the expressive nuances of a Beethoven late quartet, cannot be obtained by using words alone without missing the point. Indeed, the requisite words simply do not exist. There is the well-known anecdote that Schumann, asked by a listener what the piece he had just played was about, replied by sitting down and playing it again. All works of art are like this, which is why art could ignore the effects of science. A poem is unique; its prime essence is lost if paraphrased. A human being is unique, and we are all aware that getting to know somebody is a vastly different experience from merely reading a character sketch or, as in science, obtaining the person's vital statistics. Just as our aesthetic sensibility grows from childhood through repeated acquaintance with works of art, so does our moral sensibility through the repeated acquaintance with what is right and what is wrong. The rhetoric in literary, philosophical and religious works plays a huge part, but knowledge here is of a kind that first must be immediately and uniquely experienced by a person. It takes empirical precedence over knowledge by description, including science. In a way, it is what has survived of the world of magic, and it has survived because it is, in a real sense, more fundamental for the human spirit than science.

It is surely evident that the scope of science to inform us of the total nature of the world, including ourselves, is limited. Exactly where those limits lie in any given area will be a subject of debate – sometimes of passionate debate – even within science itself, but I think it is, sadly, not always utterly evident to everybody that, indeed, there are regions of human life where science cannot have anything relevant to say. During a visit to England in 1921 Einstein was asked by the Archbishop of Canterbury what were the implications of his theory of relativity for theology. His inevitable answer was,

'None. Relativity is a pure scientific matter and has nothing to do with religion.' Science does without God as a matter of method. Laplace's response to Napoleon's comment that his work never mentions God was that he had no need of that hypothesis (though Lagrange exclaimed that God was a beautiful hypothesis that explained many things).

Science, logically, should have no effect whatsoever on well-founded religion, yet it has. The reasons for this are not all subtle, by any means, though some of them definitely are. That part of religion founded on a creation myth and supported by biblical text was obviously vulnerable to scientific discovery. After Copernicus the Earth, and therefore man, was no longer the centre of the universe, and therefore, possibly, no longer the centre of God's attention. Giordano Bruno's vision of infinite numbers of worlds, Galileo's demonstration of the motion of Jupiter's moons, Kepler's elliptical orbit for the Earth and Newton's theory of gravitation all underscored man's insignificance. The general effect of these revelations was, no doubt, always limited to a comparatively few intellectuals, and even for many of these, a mere change of spatial location of mankind could hardly affect crucially the relationship between God and His creation. For most others the Earth remained central, so nothing had changed. It was only after the science of geology in the nineteenth century began to offer explanations of the origin of landscape, and the time-scales involved in the process of erosion, and, more strikingly, to point to the fossil record in the rocks and to a terrestrial history enormously vaster than anything suggested in the Bible, that religious belief throughout the Western world was changed irrevocably. The holy book was wrong about the history of man – might it not also be wrong about the creation of man? The appearance of Charles Darwin's book *On the Origin of Species* and the subsequent devlopment and dissemination of the theory of evolution killed once and for all, except for the most rigid fanatics, the idea that the Bible was to be read literally. The creation of man was clearly more complicated than the account in Genesis.

Yet all this could be taken to refine religion rather than weaken it. Emphasis on faith rather than reliance on dogma and revelation might be thought by many to make religion better founded and stronger. Science was valuable in identifying and cutting out the bits of religious belief that were irrelevant. But this interpretation, whatever its validity, is not the one commonly accepted. The influence of science on religion is generally seen to be almost disastrous, at least in Western society. There is little doubt that society is more secular than it has ever been, and the cause lies in the growth of science and the intrusion of technology into every walk of life. In a real sense science-and-technology has become the new religion. It is seen to be the origin of all sorts of freedom and all sorts of material goodies. More viscerally based is the belief that medical science will ultimately take away the ills of the world. Acceptance of science's involvement in the spheres of morality and society generally, and even in the interpretation of art and its evolution, follows naturally. Everything is measurable, and to see life as having a spiritual component is increasingly to appear embarrassingly out of date.

But for science to be religion is grotesque. It is far outside its rational limits. Whatever one may think about religion and its value, the consequences of the irrational aura surrounding science cannot be beneficial to society.

It is relevant to note that many scientists are deeply religious. Indeed, it would be unusual to find a scientist who did not feel a kind of awe bordering on the religious when he contemplates what science has revealed about nature. Nature is not only beautiful; it is also sublime. For Kant, beauty resided in things that are finite and limited, like a flower, whereas the sublimity was to do with powerful, limitless things, like the sea or the stars. It is the deep awareness of the power and limitlessness of nature that evokes in scientists that mix of ecstasy and humility commonly associated with mystical religious experiences. But the fact that science permeates practically all aspects of life these days through its technologies, and that it is institutionally Godless, must inevitably have contributed, via some sort of osmosis, to the increasing secularity of society. In the same way, its orthogonality to aesthetic and moral values has, no doubt, added to the process of human devaluation. The effect is utterly illogical and literally unnecessary, but, given our human frailty, very natural. If science attracts criticism for contributing to the spiritual ills of society as well as to some of its physical ills, it is time to emphasize its limited scope so that people can see science clearly as it is.

We now turn to a rather general limitation – the fact that science does not exist in an intellectual vacuum but needs support from beyond itself.

Metascience

Three

> For no perfect discovery can be made upon a flat or a level; neither is it possible to discover the more remote and deeper parts of any science, if you stand but upon the level of the same science, and ascend not to a higher science.
>
> Francis Bacon, **The Advancement of Learning**

Science has staked out a special territory by investigating mainly the inanimate and the impersonal, often regarding the animate as a special sort of inanimate object. It seeks out the world of repeatable, public fact that is, in a sense, timeless and unchanging. In doing so, science inevitably divorces itself from the unique and the subjective and largely from the whole phenomenon of becoming. Its statements about timeless universals build up a picture in our minds that transcends individual experimentation and that seems to raise timeless universality to a higher level. Meaning is abstracted from myriad observations and used to create an all-embracing, coherent image of the universe. But the meaning of that creation is not to be found in the creation itself, but in concepts that function on a higher level. Inescapably, science requires its metascience.

But metascience can come only from the world of value and sensibility, and here we enter a world inhabited by the three ms – mysticism, magic and metaphysics – each one guaranteed to raise the hackles of those scientists, if any still exist, who insist on the pure rationality of the practice of science. Yet, the belief that a real world exists is metaphysical; the fact that we can manipulate it is inexplicable and magical; and seeing the world as a unity, amenable to being described by a grand unified theory, is mystical.

In medieval times, the mystical experience of knowledge and understanding was called experimental wisdom. The unformulated feel for how things work is not unknown today. There are those with 'green fingers' who can get things to grow that the rest of us cannot; there are craftsmen that have a feel for their work in a sense that is uncommunicable; and even today in the highest of high technology there are crystal-growers, electronic engineers, experimentalists of all descriptions, who have a flair for doing their thing, in spite of their jobs being scientifically describable. It is the way in which some men and women of sport are better than others. It is an oddity of modern life that no one thinks it strange that our best sportsmen and best sportswomen have significantly better hand–eye coordination than the rest of us, but to claim that the best scientists are in some sense similarly endowed with a talent that distinguishes them from their fellows, in this case a talent involving a remarkable nature–mind's-eye coordination, might be considered somewhat heretical. The claim would suggest that there is an art in doing science, but, surely, science is a rigorously logical subject and those who are good at it are just those who have logical minds. A logical mind is definitely a necessary attribute, but it is not sufficient. If an outstanding scientist happens to be a public figure, he is often labelled a genius, which carries implications above and beyond any straightforward application of rationality.

The nature of flair, talent, genius is not easy to describe, and a certain degree of mysticality continues to be attached to it, particularly as it cannot be taught. On the other hand, mystical experiences about the cosmos *can* be rationalized; this is what metaphysics is about. Metaphysics builds a rational picture of the world on the basis of the Principle of Sufficient Reason – things that happen have a rational explanation. A succession of events, for example, may occur by chance, or out of necessity. If the latter, a principle of causality is applicable. Matter may be fundamentally corpuscular plus void, or it may be some sort of plenum, particle-like or field-like. What of a thing are its primary (timeless) properties, and what its secondary (variable) properties? How far can logic apply in a world so evidently metalogical? How far mathematics? And so on. If mysticism suffuses magic and stimulates metaphysical speculation, metaphysics suffuses science and stimulates its search for understanding.

The most down-to-earth evidence of this is that all of us in science are scientific realists in practice. Acceptable theories are those that are coherent with other theories and that correspond with the facts. In everyday discussion, electrons and electromagnetic fields are real entities that have an existence in a real world independent of our minds and our imagination. We believe in the existence of a real world quite separate from ourselves, about which we learn a little more with every experiment. This is the fundamental metaphysics that suffuses science in the laboratory.

And yet it is logically quite indefensible, as many philosophers of science tirelessly continue to tell us. Even scientists themselves warn against naive realism. Ptolemy, over 1,800 years ago, regarded the epicycle theory of the solar system as something that 'saved the phenomena' in the sense of giving an accurate account of planetary motion. Today, we scientific realists would regard that theory as inadequate, and would prefer the heliocentric theory of Aristarchus of Samos, even though at the time it was less accurate. We believe that the Earth goes around the Sun, not the other way about, and it is a belief that reaches the degree of certainty in the light of the theories of Kepler and Newton. Yet, Newton himself warned that there was no physical explanation of his magical action-at-a-distance gravitational attraction. From Pierre Gassendi in the seventeenth century down to Niels Bohr in the present century, examples proliferate of the view that theories are no more than instruments of prediction, convenient fictions, that do not describe real things. For positivists like Carnap and others in the Vienna Circle, one theory was as good as another, provided criteria for verification were clearly defined. Nor does Popper's replacement of verification with falsification change things fundamentally,[1] although it certainly helps to get rid of a lot of pseudo-science. Taking things to extremes, Feyerabend pointed out that 'anything goes'.[2]

These viewpoints deny the possibility that we may know what nature is really like. Whatever is out there beyond our sense organs is basically and fundamentally unknowable. All we can do is to construct models and maps and metaphors that function analogously to nature. Scientific theories are inventions of our minds with no *direct* correspondence with whatever 'reality' is. Since our theories cannot be logically verified, their status is absolutely uncertain. Therefore, in the last resort, anything goes.

The extreme anti-realist view is that if there is no one objective absolute theory possible, science becomes an art form. In denying the arational claim of the realist that man is of nature and therefore can know nature by empathy, or by *a priori* intuition (as Kant held), or by direct perception of a Platonic form, such as causality, the anti-realists convert science into a kind of metaphysical entertainment to be judged, presumably, on aesthetic principles taken from some other metaphysical standpoint.

Clearly, the meaning of scientific theories lies not in science itself, but in a metascience of conflicting metaphysical systems. One does not have to believe in an extreme metaphysical realism that claims the existence of one objective absolute theory of nature in order to be a robust realist. The scientific realist is supported by the belief in some sort of knowability of nature, that there exists some sort of sympathetic correspondence between what goes on in the physical world and what can be known by us. Scientific realism is essentially an acceptance of the old Pythagorean creed of microcosmos–macrocosmos, but with knowledge through empathy sought by experience. Anti-realists, distressed by that belief in occult correspondence, abandon the idea of the possibility of real knowledge of the world, and search for alternative criteria for judging the worth of a scientific theory. This leads inevitably to pure sophistry, to some Protagorean formula or other, in which man, not nature, becomes the measure.

The denial of a knowable reality is the burden of all such scepticism. One's imagination is forbidden to contemplate the possibility that nature is truly available as a meaningful object of knowledge. Putting it that way highlights the emptiness of this scepticism – for how absurd to legislate for the imagination! One of the laws of metascience is, surely, *that an idea that limits man's scope for acquiring knowledge runs the risk of extinction through being surpassed by an idea that does not, since the first forbids, and the other frees the imagination.*

The idea of knowledge by empathy sought by experience suggests a world of transcendental unity in which the apparent analytical divisibility into things plus their mutual interaction is a crude image. There is a oneness about the world of which we are a part, hence the possibility of empathy. Physics, in fact, is permeated with the sense of oneness. It is a sense, rooted in Pythagoreanism, that drew empirical strength from Newton's achievement: the laws governing the fall of apples also governed the motion of the planets; the ideas of force and inertia encompass all mechanical phenomena. There was further evidence of unity in the discovery that magnetism, electrostatics, current electricity and light were found to be manifestations of one aspect of nature, electromagnetism. Now, physics is positively motivated by the sense of oneness.

But matter seems stubbornly to oppose this quest. The Standard Model of the elementary particles identifies no fewer than 12 fundamental particles – 6 types of quark and 6 types of leptons – and 4 fundamental forces – gravity, electromagnetism, the electroweak interaction and the strong interaction – each of these with its own particles – the graviton, the photon, $W^{\pm}$ and Z^0 particles for the electroweak force, and 6 gluons for the strong nuclear force. There is also the yet-to-be-discovered Higgs boson, which is needed to account for the particle masses. That makes 24, and then there are the antiparticles. The physical world is made up of 48 things.

The weak interaction, the interaction that mediates radioactivity, can now be connected with electromagnetism. At present the strong interaction, the force that contains the particles inside the nucleus of an atom, and gravitation lie outside the scheme, but intense theoretical effort in the field of string theory, in which particles look like bits of string rather than points, aims to unify all the particles of the Standard Model, plus the many that arise from symmetrizing fermions (particles with half-integral spins) and bosons (particles with integral spins) at enormously high energies. It will, no doubt, given the versatility of mathematics, eventually lead to a Theory of Everything.[3] The Quest for Unity, as ancient as deification of The One by Pythagoras, is the driving programme of fundamental physics today. But calling a particular research programme in physics the search for a Theory of Everything is indicative of a mind-set that leads easily into scientism. It is, of course, not a Theory of Everything, but as a theory of matter that aims to embrace all known particles and fields at all energies it is certainly exciting and ambitious.

One consequence of this quest has been the merging of two branches of enquiry, namely, high-energy physics and cosmology. The range of energies characteristic of the various fundamental interactions is enormous, covering some 40 powers of ten. At the low-energy end is gravitation, and at the high is the strong interaction. This is not, understandably, a good situation for Unification – the disparity in energy is too much. The solution is to have everything at the same energy, something that is conceivable in the context of the Big Bang, hence the connection with cosmology. In the beginning was perfect symmetry at enormously high energies where everything was like everything else (that is, if we could speak of a beginning, which we cannot because time is mixed with space because of gravity). The subsequent evolution of the universe broke the symmetry (as it does in the Book of Genesis with the creation of Adam and Eve) and the various families of particles with their interactions became distinct (Fall from Grace). A new story of creation is being formed, but how much science and how much myth remains to be seen. It is not clear, even to the proponents of string theory, how adequately its predictions can be tested empirically.

The search for Oneness is essentially a search for an ontology – a being, rather than a becoming. As such it is theological in character. Nature appears not to recognize the distinction as she ought, at least, according to the mathematics employed. There is this confounded Heraclitean change. If we look at the world as a whole there can be no concept of change. 'I am that I am', or, an inscription on the Temple of Isis found in a footnote to Kant's *Critique of Judgement* that Beethoven kept under glass on his work table:

> I am all that is and that was and that shall be, and no mortal hath lifted my veil.

Something as monistic as that could certainly have its veil lifted by any modern-day string theorist, who may not stop at the veil! But a moving target is more difficult to hit. Moving with respect to what? And thereby hangs a tale. We are familiar with motion relative to something that is stationary, but is all motion like that? Can there be such a thing as absolute motion? Or is all motion, all change, bound to be relative?

A suggestive monistic idea is that of a universal equation of motion, found in the *Feynman Lectures on Physics:*[4]

$$U = 0.$$

On the left-hand side of this equation U incorporates all the laws of motion of mechanics, electromagnetism, etc. This equation looks pretty, but, as Feynman points out, it contains no more than the individual laws of mechanics, electromagnetism, etc. The monistic nature is only apparent, something that would be immediately evident as soon as one applied the equation to a particular case. There can never be laws of change with unlimited context. A duality must always exist between the system studied and the rest of the universe, and this dichotomy is always, to some extent, a matter of judgement. Bohm emphasizes 'the qualitative infinity of nature', a modern version of Plato's Principle of Plenitude. No changing system can be universal since it always must work with a finite number of kinds of things and leave out the rest, which brings into being the bothersome business of boundary conditions at the interface of the system and the rest of the universe. Inevitably this involves a dichotomy, a contingent division between the describable, causally related events of the system and the 'state' or 'random' fluctuations of the environment. No chance of an all-embracing theory of motion here. It seems that the idea of Oneness in science may be applicable to being but not to becoming.

But that is no different from classical monism. Thales saw the world as a single substance, water, and his successors differed only in the quality of the universal material. The most beautiful exposition of this idea was that of Spinoza. The real nature of substance was that it was a conception formally independent of the concept of any other thing. Thus, nothing whose attributes are the attributes of outside causes can be called a substance. Two substances which have different attributes can therefore have nothing in common and so one cannot be, in any sense, caused by the other. If more than one substance existed it would require an explanation, and that would introduce elements other than the substances themselves. Therefore, there can be, logically, only one substance. Spinoza called this substance God or nature.[5] This logical pantheism is the rational conclusion of taking 'substance' as the defining element of monism. String theory appears to be aiming at a mathematical definition of the Spinozan substance.

A more abstract monism assumes that the essence of a thing consists of its relations with everything else. Thus, a book on the table is different from the book on the floor. The reality of the universe is that of an interrelated unity. Platonic forms and scientific laws are merely abstract universals that are nothing without instances, but, conversely, all instances are nothing without universals. As Kant put it, thought without content is empty; intuitions without concepts are blind. Universals and particulars are distinguishable but not separable aspects; only the whole, containing both aspects, is real. Science would go along with that.

But then it seems to follow that the highest degree of reality is therefore the Absolute Whole, and Hegel argues that since the whole precedes the distinction into knowing mind and object known, the whole is the whole of knowledge. The reality of the world is then Absolute Idea.[6] Add to that the logical monism of a strand in modern philosophy in which logical relations link the world into a whole, and one has a fair description of the world of present-day mathematical physics.

One vision of monism that does not appeal to science is the idea that things are related organically. Science spent a lot of time killing off the world, so it is not surprising that this view would be regarded with horror in the unlikely event of its being noticed. It was, nevertheless, promulgated in the early part of the twentieth century by an extremely distinguished mathematician, Alfred North Whitehead, who advocated the abandonment of materialism in favour of the doctrine of organism.[7] In his view the distinguishing features of the universe are like organs in a living being, functioning according to a grand pattern and organized into a purposeful structure. The whole concept of materialism only applies to abstract entities, but the concrete enduring entities are organisms and these determine the actions of molecules, atoms and electrons.

> In this theory, the molecules may blindly run in accordance with the general laws, but the molecules differ in their intrinsic characters according to the general organic plans of the situations in which they find themselves.

An electron within a living body is different from an electron outside it, by reason of the plan of the body. In analysing parts, science inevitably misses the intrinsic organic relations that exist, and understandably sees no evidence of purpose. Whitehead reintroduces teleology into the universe and thereby reconnects the link between the natural world and the world of value, a link long abandoned by science.

Whereas substantive and relational monism has found a central place in modern physics, the idea of organic mechanism has not. The nearest influence is perhaps seen in the so-called Anthropic Principle, whereby the universe is seen to be constructed so that human life is possible.[8] But as Einstein wrote in a letter to Max Born:

> Living matter and clarity are opposites – they run away from one another.[9]

If the world is indeed an organic unity and to that extent alive, then it seems that we cannot hope to achieve mathematical clarity in our understanding of parts of it.

If the world is not a unity of some sort, then the theories of science need have only specialist application and are not required to be coherent with one another. Few in present-day physics believe the latter. For most, it is entirely natural to suppose that a unified view of the world is possible, and, moreover, it would be a view that would correspond with reality, and not in the least arbitrary. Unfortunately, however, the quest for unity has led, in recent times, to the grandest pluralism ever conceived, that of Many Universes, which is Plato's Principle of Plenitude writ large.

The need for Many Universes stems ultimately from the fundamental statistical nature of the quantum world, and this manifests itself in three main ways. In quantum theory itself, the assignment of reality to the wavefunction in the Many Worlds interpretation means that the wavefunction never collapses, but that the world continually branches into myriad possibilities. In the search for a unified theory, the existence of irreducible uncertainties that are present fundamentally in nature means that any unified theory must be statistical in character. As mentioned before, this implies that we must consider, according to Stephen Hawking, 'an ensemble of universes with some probability distribution'.[10] Cosmology has the problem of accounting for the observed features of the universe and, in particular, how these features evolved from the Big Bang. The uniformity and isotropy of the universe are peculiarly troublesome features given the prevalence of quantum fluctuations in the beginning. One solution according to Alan Guth is the Inflationary Theory, in which one bit of the universe expanded so rapidly that the effect of fluctuations was diluted, and it is this bit of the Universe with a capital letter that we call our universe.[11] Our universe is therefore only one member of a meta-universe that contains countless inflationary bubbles for ever isolated from each other.

The confident extrapolation from laboratory discovery to the whole universe and to the creation itself is striking evidence of a profound faith in the uniformity of nature, and in its knowability. But in many cases it works. It is also testing the principles of materialism to their limits – theories are still about glorified billiard-balls in motion, though the billiard-balls be fuzzy and their motion unpredictable. The existence of life and its organic nature is an ever-present fact that is disturbing. At the very least it suggests that a more sophisticated materialism is called for. The creation of a higher level of materialism would certainly affect our view of the cosmos, so it is perhaps premature to focus so much attention on a cosmological Theory of Everything before the problem of life itself is tackled in a more sophisticated way than it is at present.

The perspective of the so-called 'soft sciences' – sociology, psychology, cognitive science, etc. – tends to be that, one day, they will become 'hard sciences', like physics or chemistry. Physics, the most fundamental, is taken as the paradigm. But physics has come a long way from its classical, mechanistic roots. Mechanical determinism has become seriously qualified. Moreover, the subject has become much less homogeneous than it is commonly perceived to be. The nature of the paradigm has changed and is changing. New perspectives even in physics are to be found.

A perspective is a point of view. Can one have a point of view in physics? More precisely, can one have a point of view other than the point of view? What is the point of view in physics? I suppose it is the conception that there is a real world out there and that truths about that real world are knowable through the empirical methods of science. It is a perspective that has proved to be remarkably powerful, infinitely more so than, for example, the perspective, once a serious competitor, that saw the world as a magical realm full of occult and demonic forces. That is to say, it has proved to be powerful for engineering and for providing a rational picture of the universe, but not for foretelling one's future or for casting spells or for stimulating any sort of spiritual imagination. This failure to satisfy some of the deeper needs of humanity is why the disreputable magical perspective lives on alongside the scientific one, evidenced by tabloid astrology and the thousand superstitions that mind is heir to. Holding both the magical and scientific perspectives quite seriously was certainly possible for Newton and his contemporaries, but it would be rare, to say the least, to find this mind-set exhibited by any physicist today. But replace the magical with the religious perspective, and there are plenty of scientists who hold the scientific perspective alongside the religious perspective, the one to do with scientific truths and the other with religious truths. It is therefore not too optimistic to foresee that, one day, truths about art and the humanities will be found compatible with and possibly illuminated by some of the ethos that underlay that branch of magic known as natural magic.

Each of us as a human animal carries a mental portfolio of perspectives, every perspective motivated by some interest or other. We are generally interested in understanding the physical world, in the behaviour of our fellows, in art and in God; and, as a consequence, there exist the scientific, the ethical, the aesthetical and the religious perspectives. Our aim is to discover interesting scientific truths, interesting ethical truths, interesting aesthetical truths and interesting religious truths. Evidently there is no such thing as Truth with a capital T without any subscript; there is only truth$_{sci}$, truth$_{eth}$, etc. (This is nothing more than Nietzsche said long ago, and essentially reiterated by Wittgenstein.) It is a sad reflection on much of today's general scientific literature that the multiple nature of truth is ignored. But God is certainly a problem. There is this awkward thought that somehow this universe of ours came into being some 10^{10} years ago, and that suggests the need for a Creator, God, The Prime Mover. Cosmology has taken up the implied challenge with considerable flair. God, the Creator, can be dispensed with, according to Stephen Hawking, by simply eliminating the beginning. Quite apart from the God problem, any mathematical theory of the universe with a whacking big singularity in space-time that the beginning introduces is not to be borne. If somehow a mathematical framework is found that combines quantum theory and gravitation and that circumvents this problem of the singularity, so we have in some sense a world without beginning or end, then we might begin to ask questions like why we and the universe exist. Hawking continues, 'If we find the answer to that, it would be the ultimate triumph of human reason – for then we would know the mind of God.' This jump from 'how' to 'why' is as unsafe as the jump from 'is' to 'ought' scorned by Hume. In any case, people won't wait for quantum cosmology to sort out its mathematical problems before asking 'why', nor do they.

It has been said that science will ultimately eliminate the need for God; science will develop computers indistinguishable from people; science shows that we are nothing but biochemical machines; and so on. Out of this comes the pernicious religion of scientism, as dogmatic as any other religion, holding that only scientific truths count as truth. It is acutely embarrassing to come across this sort of thing in the writings of some of our best intellects, though, in this respect, there is little to choose between those who promulgate such a view and those in the humanities who espouse bizarre anti-realist views.

One can see only too clearly where scientism comes from.

The religious perspective, far from improving man's lot has given rise to wars, persecution and horrible cruelty, and seems set to continue this record. For all the study in aesthetics it seems always to come down to 'some people like Beethoven, and some people don't', and for all the study in ethics we still behave as well or badly as ever. But science is different. It has changed the power of humanity to control its physical, chemical and genetic environment absolutely and irrevocably. In that immensely important sense, scientific truths are the most powerful, and it is not too much of an extrapolation to claim that they are the only truths available to us, or worth bothering about.

But to make that claim is to forget at least one crucial point which is that the whole scientific activity of the world is founded on scientists being honest. In other words, the scientific method as practised involves, among all sorts of non-scientific things — intuition, imagination, a feeling for elegance and form — questions of morality. Ethical truths enter. There may be a perfectly good scientific reason for ignoring that point lying well off the straight line, but it had better be rationally defensible and not motivated by personal ambition. Fraud in science is anathema. But so is the emotional prespective that sees the universe as purely scientific. However powerful scientific truths are, it remains a fact that people, including scientists also, view the world from the points of view of ethics, aesthetics, religion and, yes, sometimes even magic, and much else besides.

That eclecticism of viewpoint permeates science itself. To speak of the Scientific Perspective is to speak very roughly indeed. Nobody supposes that the interests and points of view of the psychologist coincide with the interests and points of view of the geologist, though it can be confidently taken as read that they share the same scientific ethic and approach. To digress a moment, another feature that they share is that neither claims to see the world at large as solely his domain, for example, as the world as mind or as plate tectonics. They differ, of course, in that geology has no claims to illuminate the nature of the human condition whereas psychology has perfectly legitimate claims, if only to quantify things like reaction times and other measurable quantities. On the other hand, there are some microbiologists who claim that the nature of man is the nature of a molecular automaton. To describe some of the functions of an animal in terms of chemistry is to adopt a perfectly good working perspective, and if a physicist interests himself in this field by adding some quantum behaviour, that too is a good working perspective. But the addition of quantum theory, in itself, in no way justifies extravagant extrapolations. Macroscopic matter is basically billiard-ball stuff even if the billiard-balls behave more fuzzily, and billiard-ball descriptions remain billiard-ball descriptions, for ever constrained by the language of the perspective. It cannot be other than true that a human being is a molecular automaton because the truth is tautological, being a microbiological truth. But that is not all a human being is. Another truth, this time from physics, is that a human consists of electrons, quarks, gluons and photons. In the context of, say, the study of mind, this is probably the most boring truth in the world.

Within physics itself the situation is more subtle. This brings us back to the question of the possibility of different viewpoints even within specialized topics like physics, and the answer is, of course, yes, it is possible. Indeed, the more experience of physics in action one has, the more difficult it becomes not to suppose that the number of different viewpoints equals the number of physicists. If there is a major divide, it is between experimentalists and theoreticians. Crudely speaking, the experimentalist has a perspective that starts from his equipment and looks for something to measure, and the theoretician starts from his mathematics and looks for someone's equipment. Happily these two perspectives complement one another. The experimentalist would seem to have the narrower perspective. Having built and commissioned equipment, which these days is bound to be fairly sophisticated and therefore expensive, he naturally wishes to exploit it to the full, but this means that having a crystal-growing machine he is stuck with growing crystals, or having a spectroscopy machine he is stuck with spectroscopy. But the perspective of theorists, in principle of infinite breadth, can be just as narrow. Having mastered some mathematical technique the theorist looks for problems where it can be applied, or, more commonly following the advent of powerful computers, having developed a Monte Carlo program, say, he looks to use it as widely as possible.

In practice, the different perspectives of the experimentalist and theoretician arise inevitably from the need to specialize, since there are few physicists today, and arguably none, who can claim a general expertise in physics. Of more interest are the different conceptual perspectives that exist, since these touch on fundamental issues.

What does it mean to understand some familiar physical process or system? Does understanding consist merely of knowing the fundamental equations of physics, from which, in principle, quantitative descriptions of the physical processes or systems can be deduced? Or, that sort of understanding by any physicist being taken for granted, does real interest only come from following the details of such a deduction? If so, does a computer-generated solution constitute real understanding as much as an analytic solution? From the perspective of the computational physicist the real understanding is embodied in the writing of the program using the appropriate equations; from the perspective of the analytical physicist real understanding comes from seeing the quantitative strands that stem from those equations and constitute the solution. There are two perspectives operating here, one content to limit understanding to a program that can be applied to a number of specific systems, the other content to obtain a general understanding of all such systems, usually at the expense of numerical accuracy. There are, once more, two complementary perspectives, but they can be, and often are, combined fruitfully.

But this is not the case when the complexity of the problem is so intense that it can apparently be handled only by numerical computation. More and more interesting problems are like this; for example, many-body processes in solid state, instabilities in magnetohydrodynamics and meteorological modelling. Perhaps most remarkable is the calculation of hadronic masses, taking years on a specially designed computer. What is the character of understanding in these cases? Increasingly it seems to be an irrelevant concept, inapplicable once the program has been written and fed into the computer and the machine has successfully halted. A perspective in physics that engenders this view is significantly different from the traditional one of seeking a deep understanding of the physical world. The latter should not be abandoned without a fight, no matter how complex the situation, but no doubt this view will be branded as desperately old-fashioned by those with this numerical perspective. But does not 'understanding' involve relating, however approximately, to our immediate experience of the physical world? If this means attempting to translate into crude, approximate models or analogies, this is going to be better than 'Here are the numerical results – take them or leave them!'

There is also a mathematical perspective – the view that the equations of mathematical physics are primary in some sense, as distinct from being elegant, operational and mutually coherent summaries of observed data. The supreme example of this is to be found in the continuing controversies surrounding the so-called collapse of the wavefunction in quantum theory. Schrödinger's equation is taken to be embodied in nature like a Platonic form, and its solutions, regarded as having a reality of their own, cannot just disappear in a measurement. From this perspective are generated a number of interpretations – the Many Worlds Theory, the Many Minds Theory, the Coarse-grained and Fine-grained Histories, and several others – that can appear simply fantastical. More conservative perspectives generate other explanations – the introduction of a stochastic element into the Schrödinger equation, the speculation that gravitational effects enter, or simply that quantum particles follow classical trajectories in a special quantum field. Or, going back to Bohr, that the Schrödinger equation is merely an instrument for predicting the results of measurements of quantum systems and has no more reality than that. Of course, none of the truths of nature that are claimed by these perspectives conflict (so far) with observation. It seems that, more and more, some physics is becoming a matter of taste to be judged within the discipline of aesthetics.

This conclusion concerning aesthetics may well apply to present-day cosmology in its speculations regarding the first few moments of the Big Bang and the closely allied attempts

Metascience

to describe elementary particles at energies far beyond what present-day governments agree to fund. The interest in extrapolating from theories founded on observations in the laboratory, astronomical and otherwise, to processes far beyond experience has paid off handsomely regarding our understanding of observable things like the relative abundance of the elements, stellar processes and the cosmic microwave background. The success of such bold extrapolation is what motivates the principal feature of this cosmic perspective which sees physics as providing the definitive genesis and eschatology of the universe. Or rather, since the universe is a unique thing and no science can deal with a unique thing, the aim is to produce definitive genesis and eschatology of some statistical ensemble of universes. This perspective already suggests the existence of truths of such a remote nature that interest in them may be severely limited to the coterie that breeds them.

There are several points to be made. Certainly, physics as practised is not the monolithic thing it is often thought to be. Different perspectives, founded on the different interests of physicists, abound, and each perspective generates a particular way of seeing the physical world. A discussion about some problems in physics taking place between two physicists holding different perspectives can be amazingly difficult. To one, something is intuitively obvious; to the other the something needs to be established formally – and so on. I speak as a physicist here, but I would be surprised if this were not recognizable in other sciences. Nevertheless, these subjective elements become subsumed within the Scientific Perspective which is motivated by the interest of all scientists to discover, objectively and publicly, truths about the physical world. In the end, any discussion must come down to 'OK, what experiment do we set up 9 o'clock Monday morning to test this?', or something equivalent. Physics is, after all, an empirical subject.

The truths that science discovers are scientific truths, and there are many other sorts of truth that people are interested in. Elevating scientific truth to Truth with a capital T is scientism. Even if scientism is avoided, there is an idealistic view of science, often held when we are young, that worships all scientific truths as holy. Experience of active research involving persuading bosses, committees, funding agencies, etc. reveals all sorts of heresy, the worst being that some scientific truths are held to be of no interest or, worse, pernicious. Clearly, non-scientific perspectives are at work here. Political, social, financial considerations inevitably enter, sometimes steering the thrust of research in directions other than science would choose. Science is not practised in a human vacuum. The research laboratory is not a monastery. Values outside science enter into its daily practice.

Ever since the discovery of the quantum world and the observations in astronomy that has made cosmology possible, physics has needed new perspectives of nature. Mainstream physics pursues an essentially classical investigation of more and more complex systems of matter and radiation using traditional perspectives. Though deeply mysterious still, the quantum world is being manipulated successfully by following well-tested quantum-mechanical rules. Theories of complex processes are advanced and tested in the usual way by experiment. Problems in physics of a fundamental nature, such as the interaction between the quantum and classical world exemplified by the 'collapse of the wavefunction' and the rôle of gravitation in quantum theory, can usually be ignored. Much of the activity is directly motivated by application, optimistically for the benefit of humanity in the Baconian tradition. Normal perspectives are adequate. But when one encounters the speculations of physicists about the quantum realm, the origin of the universe, the mind, the sentience of computers, immortality, and other topics of an undeniably fascinating character, one feels the lack of a useful perspective. What are advanced are not obviously testable empirically, but they do not contradict any physical principles. They are advanced and discussed in the usual rational manner. They are ideas about the nature of things that seem to transcend normal scientific practice.

There is clearly a new perspective in physics at large. But its nature is distinctly problematical. Look, for example, at the problem of the isotropy of the universe. Observation shows that the cosmic background microwave radiation is remarkably isotropic (even when the earth's motion is taken into account) with only very small fluctuations. Given a quantum-mechanical universe one would expect much bigger fluctuations. There is a choice of theory here – does that mean that isotropy is a fundamental feature of the universe dating from the Big Bang, or is there some physical principle that engenders isotropy no matter how anisotropic the Big Bang was? The former theory is a killer as far as application for funding is concerned; the latter is open-ended, and therefore more attractive to our problematic perspective. It has to be said that the theory that there is some principle that operates what seems to be, in a certain view, a highly improbable isotropy has the sort of vector, or direction, that opens the possibility of discovering new things. It is still . a perspective that is problematical; it assumes that this unique thing, our universe, could be otherwise. But can something unique be otherwise? Hardly, but our understanding of its nature can certainly be otherwise.

The discovery of the electron at the end of the nineteenth century and of the photon at the beginning of the twentieth affected all of physics. The new quantum theory successfully described the structure of the atom and transformed our understanding of all forms of matter. The studies of solids, liquids, gases and plasmas were all revolutionized. What was fundamental in physics affected directly the whole of physics. A single perspective held sway. This is not the case today. Once the existence was established of elementary particles other than the electron, the proton, the neutron and the photon, the new topic of particle physics began branching away towards its own particular vision and developing its own particular perspective. Its new discoveries ceased to be influential in physics as a whole. They did, however, find an increasing relevance in another topic that was busily sprouting away from the main trunk of physics, namely, cosmology. Founded on Einstein's General Theory and succoured by the rich observational discoveries of astronomy, it, too, began a life of its own, with a perspective and aims that sometimes seem more akin to those of theology than those of physics. A certain messianic zeal became evident in both off-shoots, the one intent on finding a Theory of Everything, the other on explaining why the universe could not be otherwise. Neither offered much hope that any material benefit would emerge from these researches, unlike mainstream physics. On the other hand they both offered, and still offer, a sort of spiritual benefit (unlike mainstream physics!) in the form of a coherent universal model of the world, and they compete in this respect with the more familiar purveyors of spiritual benefit, religion and art. However that might be, it seems increasingly evident that mainstream physics and cosmology/particle physics have less and less to do with one another. With such different perspectives they deserve different names – physics and hyperphysics, perhaps. Some change in nomenclature seems overdue. It may not be now. But it will come. The readiness is all.

Four

That the sense of man carrieth a resemblance with the sun, which (as we see) openeth and revealeth all the terrestrial globe; but then again it obscureth and concealeth the stars and celestial globe; so doth the sense discover natural things, but it darkeneth and shutteth up divine.

Francis Bacon, **The Advancement of Learning**

(quotation from 'One of Plato's School')

But how far has science left magic behind? Scientists are, believe it or not, human, and often motivated by passions that are far from being scientific. If science has left the heady world of magic far behind, it is not as evident that scientists have done the same thing. Certainly, early scientists like Gilbert and Newton were firmly in the magical realm with beliefs that were far from what we would now recognize as scientific. We imagine that today's scientist is more enlightened. But with his monistic view of the universe, is he? The world presented to him as a human being appears to be irrevocably split into a world of fact and a world of value. It seems to consist of both matter and spirit, to be part dead, part alive. It contains elements most of which lack consciousness and to support forces that are mostly impersonal. It contains conscious beings and personal forces. How can this world be understood? At one pole is science, which has made the world of unconscious and impersonal things its own, and from that power base lays claim to the whole universe. At the other pole is religion, whose conscious and personal world stands for ever opposed to that of science. In between religion and science is the vast tropical realm of human sensibility and magic in which most of our experiences belong; the magic, like the Pythagorean music of the spheres, unperceived, not because we lack fine enough senses, but because it is all around us, and we take it for granted.

Science and magic were once intermingled, and I would like to rescue magic from the aura of superstition, nonsense and sometimes evil that tends to surround it, and give it a chance to contribute to our understanding of the world we live in. I believe that some of the fundamental meaning of magic that attracted men of learning like Marsilio Ficino, Pico della Mirandola, Giordano Bruno, William Gilbert, and later, Isaac Newton and many others in the then newly born Royal Society, and indeed many across Europe, was, and is still, important for the human spirit. It was the so-called Enlightenment that elevated reason to unsustainable heights and initiated a perspective that dismissed all magic as superstitious nonsense. I believe that something important was lost thereby, and I believe that it is worth while to try to distil an interesting meaning out of magic that may serve as a bridge linking science to the humanities. Its positive virtue is its stimulation of the imagination. Science tends to be somewhat puritan in that it explores a mechanical and predictable world without the help of music, poetry and humanity. Magic reminds us that the world is not like that.[1]

Over the greater part of human history the world was considered to be alive, and man, as is his nature, sought to control and manipulate its elements using the tools of magic, prayer and sacrifice. As a living world it was inevitably capricious, but a world that shone with divinity was surely amenable to the workings of faith; one inhabited by gods able to be manipulated by placatory sacrifice, a universe of spirit by magic. Power seemed to be within the grasp of the priest and the magus; but the priest could never presume to predict the action of God or gods, nor could the magus offer greater certainty, though he had few qualms about seeking to compel gods, demons and nature. An animated world in which not only animals and plants but also sea, sky and the very rocks were alive was bound to be awkward to handle. It gradually became clear that in order to achieve an element of certain power, the world had to die and its corpse had to be procured and dissected.

In the West, from the sixteenth century onwards the physical world was quietly put down, carved and classified, and the biological world followed. Darkening and shutting up the divine has payed off handsomely. The great champion of science (science understood in its broadest sense, i.e. knowledge), was Francis Bacon. In his book *The Advancement of Learning* published in 1605, Bacon urged his contemporaries to beware of mixing science with religion, to use the evidence of their senses, to carry out experiments, and so increase the sum of knowledge for the benefit of mankind. Today, the sort of learning urged by Francis Bacon in 1605 encompasses the most transient elementary particle, the furthest reaches in time and space of the universe, the sequence of nucleotides in DNA, and even, in some sense, the idea behind what used to be the Marxist state. Through the de-animation of the world, that arcane power over nature, so sought after by the magus, has been laboriously but steadily acquired. The achievement of science is truly remarkable. Given the requisite commitment of time, endeavour and resources, anything that accords with the laws of nature can be grasped. With such tangible powers at our disposal it is not easy to see what value there was in those old beliefs about an animate nature. Certainly no amount of prayer could have produced the steam engine, no amount of incantation evoke the silicon chip, no amount of mysticism the laser. If the passing-away of the divine world is mourned, it is more likely to be for the social cohesion it fostered than for its divine attributes. Clearly, the perception of an animate component in the nature of things confused categories and clouded man's understanding of the world in which he lived. The study of God and the study of His works had to be consciously distinguished, and Bacon admonished those who would advance knowledge

> that they do not unwisely mingle or confound these learnings together.[2]

Many who lived before Bacon knew that as self-evident truth. Pigeon-holing the religious in order to get on with the interesting activities of life is anything but new. Its articulation is as old as the atomists Democritus and Epicurus and the sophist Protagoras, and that antiquity speaks unambiguously of its usefulness, as scientists have found. Banning the Almighty from the laboratory keeps Him out of mischief. If Bacon, man of affairs that he was, inclined naturally to a pigeon-holing polychotomy, i.e. the cutting up into many bits and filing in separate compartments, he has had many precursors and many followers down to the present time. For some the physical world was ever dead, and better so. For others, doubtless the majority, the physical world could be what it liked as long as it did not interfere with their ambitions.

After four centuries of science it is perhaps difficult to capture the true flavour of the idea of the world as a living organism, but this idea was central to all of Greek philosophy.

Both Plato and Aristotle saw the stars and planets as animal, with gods existing as a category alongside birds, fish and land beasts. Even Leucippus and Democritus, though tending towards materialism, recognized soul, albeit soul made of atoms. Today's idea of the earth as a gigantic self-regulating organism, the Gaia Hypothesis, is a pale image of this ancient idea.

But if science has rejected the world as living, it certainly has not rejected the equally pervasive Greek thought of Oneness, something that permeated things and was everywhere, though appearance suggested otherwise. For Thales it was water, for Anaximander something divine, for Anaximines air, for Anaxagoras mind (nous), but the most relevant to today's science was the idea of the Pythagoreans, who saw the world as number, an idea that Plato developed into his theory of forms. Form, particularly mathematical form, is the eternal reality, and since all forms are worthy of embodiment in the world, there are boundless numbers of them – the Principle of Plenitude, as Lovejoy calls it.[3] So there is the fascinating thought that the world is replete with mathematical forms waiting to be found. Surely non-Euclidean geometry existed in some sense before being discovered in the nineteenth century. The idea of Platonic forms existing in their own right alongside the other objects of the world cannot be other than a stimulating one to many mathematicians. There is no doubt whatsoever that the idea of mathematical forms, whether Platonic or not, pervading the whole world is central to our science today. This Pythagorean legacy has been remarkably effective. It has led to the concept in quantum mechanics of the wavefunction of the whole universe that mathematically describes the undivided wholeness of the world. In this, quantum theory has become part of a tradition reaching back to the Greeks. It was a tradition of attempting to understand the occult facet of the cosmos through some unifying principle, which itself sprang from a mystical strand deriving from Orphism, which was the belief in the unseen unity of God and seeing the visible world as false. But neither member of the Greek duality – hylozoism, the world as material and animal, and idealism, the world as idea and animal – has survived intact. Certainly, Spinoza would have none of it. For him, the world was one substance. In modern science the world is material *and* idea, which is an intrinsically Spinozan concept.

A radically different element of Greek thought was that of Heraclitus, and this too finds modern expression in science. Both the hylozoistic and idealistic theories are essentially ontological in that they describe a World of Being. But to Heraclitus the world was a World of Becoming; the reality of the world was continual change, and the fundamental substance was fire, which is the only thing that does not change and die.

> You cannot step twice into the same river

runs the famous Heraclitean statement. Change in an operational sense is time, but in the differential equations of mathematical physics this is not the meaning given to time; time here is a mere coordinate, like space, indifferent to sign, everything being reversible. Such equations describe a World of Being, and they would have obtained the approval of Parmenides, who criticized the elevation of Becoming over Being, arguing that the senses deceive and that behind the meaningless bustle there was a changeless reality. But there is no changeless reality. Physics recognizes this via the Second Law of Thermodynamics. Processes are almost always irreversible – a reversible system is a very special affair, not easily arranged in practice. A dropped glass that breaks does not spontaneously regenerate itself. Too much ordering has irretrievably vanished into the surroundings. The glass has irretrievably become something else or, better, a myriad of somethings else. Order must inevitably yield to disorder. Nevertheless, there are cases where order is temporarily generated out of the implacable environment, the prime example being life. The theory of evolution is, par excellence, a Theory of Becoming. Complex systems, in general, can exhibit the emergence of relatively simple forms. Time flows one way, corresponding to our direct experience, and, as Bergson has emphasized, time is not an integer sequence, pace Pythagoras, not a now-then-now-then-now, but rather a continuous present.[4] As Prigogine puts it,[5]

> living systems have a sense of the direction of time. This direction of time is one of those 'primitive concepts' –

the existence of Kant-like 'primitive concepts' being something that Bohr emphasized. However, classical and quantum mathematical physics remains stubbornly Parmenidan rather than Heraclitean. Only in the field of irreversible thermodynamics, in cosmology containing the idea of an evolving universe, and in events involving the weak interaction, does physics recognize that time actually has a direction. Otherwise, mathematical physics gets on without an arrow of time quite happily, except in the case of quantum wavefunctions, which have to collapse and lose coherence every now and again.

Although fundamental substances were timeless, the compound bits and pieces of the world of appearance could have a cyclical existence. Empedocles, famous for his four elements – earth, air, fire and water – saw these substances combining through sympathy and disrupting through antipathy in a purposeless cycle. The idea of a cosmic cycle suggested by the period for the planets to return to their former positions – Plato's Great Year of about 36,000 years – gave a sense of change being tamed to some extent. Our modern cosmic cycle – Big Bang, expansion, contraction, Big Squeeze, Big Bang, etc. (possible if the mass of the universe is not too small) would have a period of some tens of thousands of millions (10^{10}) of years.

The nature of change bothered the Greeks and it still bothers us. For Plato, change came about partly through chance and partly through necessity, whereas for the Atomists, chance did not come into it – all motion was determined. The latter view is that of classical physics – expressed by the famous quotation from Einstein:

God does not play dice.

But the advent of quantum phenomena suggests that chance may well be a fundamental ingredient of the world. Certainly at the macroscopic level of the world of classical physics, quantum events are unpredictable, but whether this unpredictability is fundamental or not depends on the interpretation of quantum theory. In the so-called Causal Interpretation pioneered by Bohm the unpredictability stems from our ignorance of the starting conditions of the motion of the particle involved in the same way that the evolution of chaotic systems is unpredictable, though all motion is entirely determined.[6] The debate about chance and necessity continues.

Aristotle emphasized another idea about change, namely, that it was purposeful. The world continually progressed towards a higher degree of form; there was a striving towards perfection; there was a cosmic hierarchy. At the bottom was the earth, the centre of the universe, ugly with imperfection, and, like all sublunary things, subject to malformation and decay. Superlunary things were composed of pure substance, moving in perfect circles, unlike motion on earth, which was rectilinear. At the top was the perfection of form, God, the Final Cause. The World of Being was a golden chain stretching unbroken down from God to the earth, each link imbued with an aspiration towards higher perfection.

Aristotle's teleological world had motion and direction. One might say that it was a Vector World, but that is a concept that has all but vanished. Our world today is largely a Scalar World – lots of motion, but no direction. If there is a direction, it is the purely material one of technological progress. Yet Aristotle's vision of the cosmos was the inspiration of generations down the centuries. His concept of the circle as the perfect heavenly motion was exploited with remarkable success by Ptolemy in his quantitative description of the motions of the planets. The mathematical intricacy of his scheme of epicycles might seem to us today, with the hindsight afforded by Copernicus and Kepler, unnecessarily complicated, and many have thought that Aristotle's ideas held back science for centuries. But to think that is to discount the effect of those ideas on the imagination which, in their extraordinary development via Neoplatonism and the Hermetic philosophy, were responsible for the semi-mystical fervour of early scientists like Gilbert, Kepler and even Newton.

In Neoplatonism, a syncretism of Plato, Aristotle and Christianity expounded by Plotinus in the third century AD, there are two new ideas, emanation and empathy. In the

Aristotelian hierarchy the essence of a higher level emanates downwards, and so the whole world is filled with the divine emanation from God. Such an idea has disappeared from our present-day culture, leaving, somewhat bathetically, only a materialistic trace perhaps in the cosmic microwave background. The idea of cosmic empathy, on the other hand, still survives in the belief that somehow we have an insight into and an understanding of the workings of nature, and that what science describes really exists in some sense. Its expression in Neoplatonism is the assertion that each being in the world contains all within itself and at the same time sees all in every other, so that everywhere there is all; the microcosm contains the macrocosm. Like a hologram, each little bit contains information about the whole. The Neoplatonic flavour is caught in a passage from Macrobius, writing in the fifth century:

> Since from the supreme God Mind arises, and from Mind, Soul, and since this in turn creates all subsequent things and fills them all with life, and since this single radiance illumines all and is reflected in each, as a single face might be reflected in many mirrors placed in series; and since all things follow in continuous succession, degenerating in sequence to the very bottom of the series, the attentive observer will discover a connection of parts, from the Supreme God down to the last dregs of things, mutually linked together and without a break. And this is Homer's golden chain, which God, he says, bade hang down from heaven to earth.

In effect, God, defined neatly in a twelfth-century book as

> sphaera infinita cuius centrum est ubique, circumferentia nusquam

> an infinite sphere of which the centre is everywhere, and its
> circumference nowhere

was immediately reachable, according to Nicholas of Cusa in the fifteenth century, without the need to go through the hierarchical levels, a very Protestant thought. And did not this make man God-like? At the very least, man had within himself the power to understand and know the world.

Below and through this philosophical system ran streams of Hellenistic magic, alchemy and astrology. Mixed with the writings reputed to derive from the Egyptian God Thoth, latinized to Hermes Trismagistus, all this made a very heady brew. All the works attributed to Hermes Trismagistus were eventually proved to have been written during the early centuries AD, and to show the influences of Neoplatonism, but when they were turning up in the fifteenth century the intellectual excitement was intense – not surprisingly, considering the fire in passages on the Mind such as occur in the Corpus Hermeticum, translated by the Florentine philosopher Ficino:

> See what power, what swiftness you possess. It is so that you
> must conceive of God; all that is, he contains within himself
> like thoughts, the world, himself, the All. Therefore unless
> you make yourself equal to God you cannot understand God:
> for the like is not intelligible save to the like. Make yourself
> grow to a greatness beyond measure, by a bound free yourself
> from the body; raise yourself above all time, become Eternity;
> then you will understand God. Believe that nothing is
> impossible for you, think yourself immortal and capable of
> understanding all, all arts, all sciences, the nature of every
> living being. Mount higher than the highest heights. Descend
> lower than the lowest depths. Draw into yourself all

sensations of everything created, fire and water, dry and moist, imagining that you are everywhere, on earth, in the sea, in the sky, that you are not yet born, in the maternal womb, adolescent, old, dead, beyond death. If you embrace in your thought all things at once, times, places, substances, qualities, quantities, you may understand God.

The evidence of recent scientific writings suggests that the fire still burns brightly.

This Neoplatonic intoxication was further fuelled by the doctrine of correspondence, which provided the theory for much of Hermetic magic: every thing and every act on one level corresponds to things and acts on higher levels. Thus gold had a magical correspondence with the Sun, quicksilver with the planet Mercury, silver with the Moon, etc. Colours too had their correspondences: orange and gold, the Sun; indigo and dark blue, Mercury; dark brown and black, Saturn; and so on. Correspondences also existed through talismans, music, number and incantation. A vast and intricate theoretical structure embracing the sublunary sphere of the Earth, the planetary gods and the high spiritual realms of God and His angels, which postulated the action of emanation as if it were a spiritual wavefunction of the universe, and which asserted the existence of forces of correspondence, provided the magus with the power to call on higher spiritual influences, to invoke angels or demons, and to understand the nature of all things.

Hermetic fervour reached its zenith around the end of the sixteenth and the beginning of the seventeenth centuries. Literature immortalized the magus in the characters of Dr Faustus in Christopher Marlowe's play and Prospero in *The Tempest*. The religion of the Hermetic philosophy was anything but orthodox, and the Roman Catholic Church made an example of one would-be magus, Giordano Bruno, who insisted on the existence of an infinity of worlds, by burning him at the stake in 1600. Nowadays we are somewhat more relaxed about people who advocate Many Worlds theories and theories entailing an ensemble of universes.

But at the time it was, no doubt, worth hiding one's Hermeticism, out of which prudent sentiment grew an invisible college, the Fraternity of the Most Noble Order of the Rosy Cross, with its claims of boundless arcane knowledge. Robert Fludd, whose cosmos represented the sophisticated pinnacle of Hermetic philosophy, wrote a defence of Rosicrucianism and dedicated other works to the Society, but was careful not to claim membership. Francis Bacon may or may not have been a member, but his account of a utopian research establishment in his *New Atlantis* is certainly suggestive.[7] Bacon envisaged a team of 36, of whom there were 12 Merchants of Light, dedicated to the acquisition of knowledge from all over the world (professors with large travel grants?), 3 Depredators who collected accounts of experiments in all books (journal editors?), 3 Mystery-men who collected experiments of all mechanical arts and liberal sciences (librarians?), 3 Pioneers who tried out new experiments (post-docs?), 3 Compilers who classified all of this in titles and tables (text-book writers?), 3 Benefactors who extracted useful information from the experiments of their fellows (theoreticians?), 3 Lamps who directed new experiments (principal investigators?), 3 Inoculators who actually carried out these experiments (research assistants?), and 3 Interpreters of Nature who raised all discoveries to a set of axioms and aphorisms (mathematical physicists?). The aims and policies were clearly stated:

The end of our foundation is the knowledge of causes,
and secret motion of things; and enlarging of the bounds
of human empire, to the effecting of all things
possible.

And this we do also: we have consultations, which of the
inventions and experiences which we have discovered shall
be published, and which not: and take an oath of secrecy for
the concealing of those we think fit to keep secret: though
some of those we do reveal sometimes to the State, and some
not.

No mention of funding! But it is interesting to observe that what Bacon took to need 36 has to be done these days with, typically, a Principal Investigator and maybe a post-doc. and a couple of graduate students. Another difference is that *everything* is published, sadly enough.

Though cool, legalistic, politically aware, but non-Pythagorean enough never to advance the virtues of mathematics, Bacon is nevertheless imbued with a passionate belief, with perhaps a strand of mysticism, that the world could be known and manipulated for the benefit of mankind. He stands downstream of cooler currents running parallel to hot flushes of Hermeticism from Leonardo da Vinci and earlier. When William Gilbert of Colchester, reporting his pioneering discoveries about the earth's magnetism in his book *De Magnete*,[8] wrote:

We consider that the whole universe is animated, and that all
globes, all the stars, and also the noble earth have been
governed since the beginnings by their own appointed souls
and have the motives of self-conservation,

Bacon rebuked him:

> So have the alchemists made a philosophy out of a few
> experiments of the furnace; and Gilbertus our countryman
> hath made a philosophy out of observations of a loadstone.

Cool strands like Bacon's, and later Galileo's, successfully countered the contemporary heat of Hermetic mysticism and initiated empirical science. The transmission of craft-mystique, of secret knowledge, of secret societies like the Masonic Order and the Fraternity of the Rosy Cross, was no longer acceptable. Nevertheless, these institutions, especially the Brotherhood of the Rosy Cross, even if nothing of the kind ever existed, consisting of men of great wisdom and knowledge, stimulated the emerging scientific imagination. The result was the formation of the scientific academies in Italy, France and England.

One of the earliest was the Naples Academia Secretorum Naturae, established as early as 1560. Galileo belonged to the Rome Accadamia dei Lincei and the Medici founded the Accademia del Cimento in Florence in 1651. In England the Royal Society was founded by Charles II in 1662 out of a society that met in Gresham College in London in 1645 under the name of the Philosophical or, significantly, Invisible, College. The Académie des Science in France followed in 1666, the Russian equivalent in 1681 and the Academia Naturae Curiosorum in Vienna in 1687. Clearly a fervour had invaded Western Europe, and it was a fervour that was to revolutionize the world. It had its source in the magico-mystical belief that man could know the universe. Tempered by rationality and applied mathematics, it was a belief that was to be sustained down to the present time. It eventually converted astrology to astronomy, alchemy to chemistry, and magic to hard-headed engineering.

But not all magic. The correspondences between man, on the one hand, and painting, music, poetry, drama, etc., on the other, may be earthbound, but they exist. The arrangement of material objects or sounds to make art undoubtedly exerts forces of a kind on the mind. Are they essentially different from magical forces? Surely not. And can science claim to be entirely emancipated from Hermetic beliefs? or from its fervour? Surely not. Is not the abiding urge to bring religion, morality and aesthetics under the explanatory powers of science a manifestation; equally, the search for a Theory of Everything and the associated attempt to account for the beginning and end of the universe?

Five

Do you really believe that the sciences would ever have originated and grown if the way had not been prepared by magicians, alchemists, astrologers and witches whose promises and pretensions first had to create a thirst, a hunger, a taste for *hidden* and *forbidden* powers?

Nietzsche, **The Gay Science**

Today, physics searches for a Theory of Everything. What is meant by that is the actually much less grandiose search for a theory that will describe all the interactions that occur between the elementary particles. It has to explain why there is a division of matter between quarks and leptons, why some obey Fermi–Dirac statistics and some Bose–Einstein statistics, why there appears to be more matter than anti-matter, and much more. It is an exploration into the most fundamental elements of nature, and is to be linked to the creation of the universe itself. It is a grand challenge, perhaps the grandest there has been in science. One might sympathize with the nomenclature, but, of course, the theory when it is found will certainly not be a Theory of Everything – nature is too immense for that. It will be the best theory of matter we have, no more, no less.

The theory we have at the moment is the Standard Model, which tells us all about the minutiae of matter. Our bodies are matter, sublunary and prone to imperfection. We are molecules, and molecules are made up of atoms and atoms are made up of electrons clouding around a nucleus made up of protons and neutrons. The number of electrons, each of which carries a unit of electric charge (whatever that is), is equal to the number of protons, each of which carries a unit of positive charge (whatever that is) so the whole atom is electrically neutral, and the neutrons, far from free-loading, are there to stop the nucleus tearing itself apart through the mutual repulsion of the protons. The electron is a light particle, one of a family of light particles called leptons. It seems to have no internal structure, but it has a spin with half a unit of angular momentum. Protons and neutrons, on the other hand, are heavy – members of a heavy group called baryons. Like electrons, each has a spin, again of half a unit of angular momentum. Particles with half-integral spin are known as fermions, because their populations obey the sort of statistics – Fermi–Dirac statistics – that allow two particles to occupy the same dynamic state only if they have opposite spins. Bombarding the nucleus reveals the presence of another family of particles called mesons. At one time it was thought that mesons provided the glue that held protons and neutrons together, but the true situation seems more complex. Mesons have integral spins and obey the sort of statistics – Bose–Einstein statistics – that allows, indeed encourages, as many particles as possible to occupy the same state. Such particles are called bosons, a familiar member being the photon, the quantum of light. Mesons, like protons and neutrons (collectively known as hadrons) but unlike electrons, are big. They have a radius of about 10^{-13}cm and they exhibit structure. They are made of quarks held together by a field whose quantum is the gluon. The quark has a half-integral spin, so it is a fermion; the gluon has an integral spin so it is a boson. In order to build all the hadrons out of quarks, they have to have some peculiar properties. For a start, they must have either ⅓ or ⅔ of a unit charge. To form a proton take two quarks, called 'up' quarks, each with a charge of +⅔, and one 'down' quark, with a charge of −⅓. The net spin is ½ and the net charge is +1, as it should be. So a proton is a (uud) particle and the neutron is a (udd) particle. Add to the list the anti-quarks with charges −⅔ and +⅓, and mesons like the negatively charged π-meson can have the structure $(\bar{u}d)$, where the bar denotes the anti-particle. It does not end there. In order to construct all the particles that are observed in high-energy colliders, quarks need to come in more than one so-called flavour. As well as up and down quarks there have to be the flavours 'charm' (c), 'strange' (s), 'truth' (t) and 'beauty' (b) (or, more prosaically, 'top' and 'bottom'). It follows that mesons consisting of $(\bar{c}c)$, in which charm is cancelled by anti-charm, can be referred to as particles having hidden charm, or those with $(\bar{b}b)$ as having hidden beauty. Where cancellation does not occur, as in mesons like $(b\bar{u})$, we may speak of particles with naked beauty (or having a bare bottom). As if all this were not enough, quarks have to be 'coloured' red, blue or green. This galaxy of metaphors is manipulated in the esoteric branch of physics known as quantum chromodynamics.

It is a delightful account of matter as we more or less know it. The Theory of Everything will tell us why the world is split into fermions and bosons and why the mass of each particle is what it is. It will get gravity into the quantum picture and how space-time may be quantized. We have some way to go before the story of matter is complete.

Yet 400 years ago there was an infinitely more wide-ranging Theory of Everything, now completely abandoned. It grew out of the philosophies of Plato and Aristotle and the magical traditions of Hermetic and cabbalistic philosophies. It explained everything of relevance to the human spirit in terms of an intricate cosmology which included supernatural beings and occult forces. It spanned religious contemplation at one pole and magical engineering at the other. It linked gods and men. But it had nothing new to say about the fundamental constitution of matter. That orthogonal theme was best left to science.

The basic questions of how to understand the world long pre-date science and are still urgent. What forces exist, and what is their nature? There are the obvious impersonal, mechanical forces, the pushes and pulls of science, that act on matter, animate or inanimate. But there are also forces that are of the mind and affect the mind, and these can be uniquely personal. We might recognize four attributes that forces can have – they can stem from conscious or unconscious entities, and they can evoke personal responses and obey impersonal laws. In the Magical Theory of Everything conscious forces and personal responses were involved in religion and human behaviour, conscious forces obeying impersonal laws figured in demonic magic, unconscious forces and personal responses informed natural or spiritual magic, and unconscious forces obeying impersonal laws eventually became the defining elements in what we now know as science (see Figure 1).

The most compelling belief we have is of our own consciousness and uniqueness. Our acts are self-evident demonstrations of forces that are conscious with evident personal consequences. We are surrounded by fellow human beings whose behaviour also betokens the action of conscious, personal interactions. A belief that God exists implies the existence of conscious forces of good, and belief in the devil implies the existence of conscious forces of evil, either of which can affect one personally. Thus, whether one believes

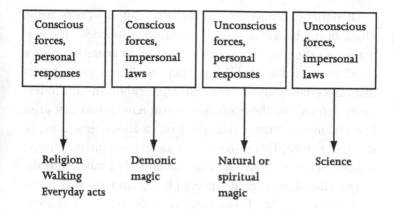

| Conscious forces, personal responses | Conscious forces, impersonal laws | Unconscious forces, personal responses | Unconscious forces, impersonal laws |

| Religion
Walking
Everyday acts | Demonic magic | Natural or spiritual magic | Science |

Figure 1

that these forces are natural and earth-bound and limited to the society of humans (and maybe to their pets), or whether one believes in supernatural beings controlling these forces on a cosmic scale, the existence of such forces in our universe is not in doubt.

If an extrapolation of the existence of consciousness beyond man is made – perhaps by invoking the Principle of Plenitude – then all sorts of spirits and demons, and ultimately God, appear. The question that arises is, are these consciousnesses above law? If they behave capriciously, without apparent regularity, then propitiation is the only way of dealing with them. But if some, at least, submit to law, then there exists the possibility of demonic magic, of invoking and commanding spirits. In order to do so it is necessary to understand the impersonal laws that demons must obey. These are the conscious forces obeying impersonal laws that are the theoretical base of demonic magic and, indeed, of witchcraft and voodoo. But we need not doubt their existence because of mumbo-jumbo.

Then there are unconscious forces that affect us personally. Think of the beneficial effects on the consciousness of a wee dram – or the more compelling effect of an anaesthetic, or a knock on the head, or of drugs. They are all impersonal forces that can certainly affect the mind. More subtle and much less easily defined are those unconscious forces, forces that arise from inanimate matter, that affect our feelings, emotions, in short, our spirit. This is the area of natural, or spiritual, magic which conceives of an order of existence that transcends the simple animal, biological, material being, and it is very much alive and well today. It has to do with the power of charms, of words, of music and of craft (see Figure 2). It is a power invested in the link of occult spiritual sympathies and antipathies that connects one thing with another. The Magical Theory of Everything has the basic tenet that the universe is permeated with spirit, cosmic and human, animate and inanimate, that hidden connections exist between all things, and that nothing is meaningless. This is what makes astrology and alchemy meaningful. It was

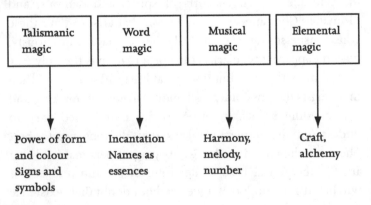

Figure 2

claimed that an understanding of these connections allowed one to lead a full spiritual life.

Unconscious forces obeying impersonal laws are the domain of science; conscious forces eliciting personal responses are the domain of religion. The conception of magic, as ordinarily understood, is less secure. The belief that God exists implies that forces of good and bad inhabit the universe. Often these forces are seen to manifest themselves in spiritual beings – angels, saints, devils, satans – which interact meaningfully with one's unique, conscious self. This is the domain of religion. But even religion has a magical flavour if it allows miracles, and if it allows the existence of individual spirits, it approaches perilously close to demonic magic. Spiritualism is clearly a form of demonic magic, though the demons be only departed loved ones. Science, of course, has nothing to do with such supernatural entities. Thus, demonic magic itself is reasonably distinctive. It shares with religion the belief in supernatural, conscious beings, and with science the belief in the existence of natural law. But in believing in supernatural beings it distances itself completely from science, and in emphasizing the practice of law to control these beings, as distinct from prayer and propitiation, it distances itself irrevocably from religion (see Figure 3).

The idea of natural, or spiritual, magic is more diffuse and a good deal more subtle than its disreputable associate, demonic magic, but it is less well defined because of its many facets. Yet its power to stimulate the imagination suggests that it has a meaningful rôle to play in today's culture. It enters many activities that have acquired individual names for themselves, and it is not always recognized as the energizing component. It has to do with the interaction between the material world and the mind, as, in fact, does science when it produces

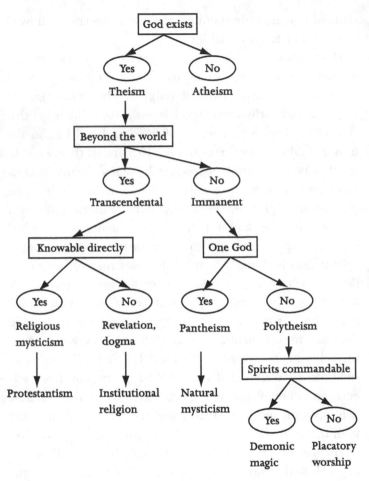

Figure 3

theories to explain phenomena; but it distinguishes itself from science by its concern with spiritual, rather than material, interactions and with its recognition that these interactions are unique to the individual. Science deals only with the general – often only with statistical ensembles – but, and it cannot be said too often, never meaningfully with the unique.

Spiritual magic recognizes the power of form and colour in the magical effect of talismans. This basically is what architecture and painting and even decorating are about. The heaven-ward aspiring spire, the form, light and shade in a painting, the harmonious disposition of shapes, colours and textures in a room – what are these if not embodiments of enchantment through the creation in each case of what, in effect, is a talisman? In the Magical Theory of Everything the talisman is crafted out of matter to attract particular spiritual powers that permeate the universe and are embodied in the planetary beings. This overlaps to some extent with demonic magic insofar as planetary spirits are believed to exist, but in natural magic, unlike demonic magic, there is no element of compulsion. The idea of planetary beings harks back to Plato's planetary animals, and further back still to Babylonian astrology. Thus, Saturn is associated with privation, Jupiter with expansivity, Mars with aggression, the Sun with vitality, Mercury with conjuring, Venus with love and peace, and the Moon with change and inconstancy. Certain forms, the signs or sigils of the individual planets, existed and could be used to attract the desired power. The circle, the triangle, the square, the cross, the swastika, the star are topologically absolute and would appear in any sign language. Such a shape inscribed on the correct metal could serve to deflect the malignant thrust of evil spirits, or bring luck, or good health. Such is the simplest talisman, and in how many cars on the road do we see modern variants! But at the highest level talismanic magic is just another way of describing the power of form in art. In other words, it is real and commonly experienced in the way that demonic magic is not.

Another aspect is word magic. Language is the prime means of communication between minds, and it has an important rôle in thinking itself. The spoken word in oratory, the written in poetry, has undoubted power and is real. Word magic claims more than just power for words: it embodies the belief that to know the 'true' name of a thing is to know its very essence. The true names of things are occult and known only by magicians, and only fully paid-up ones at that.

> In the beginning was the Word, and the Word was with God, and the Word was God.
>
> [John 1.1]

The idea of sacred words certainly pre-dates the Magical Theory of Everything. Each of these special words, in some occult sense, is the thing to which it refers; it contains the essence of the thing in a kind of intimate organic sympathy, so that to utter the word is to possess, control or merge into the thing itself. One of the most sacred examples of this conception is found in the Hebrew cabbala, in which God is the Tetragrammaton IHVH, an unpronounceable word whose nearest translation is Jehova. The Christian cabbalists, for example Robert Fludd in the seventeenth century, employed considerable finesse and ingenuity to accommodate the four-fold structure of the Tetragrammaton inside the doctrine of the Holy Trinity: the letters Yod (I), He (H) and Vau (V) were interpreted as the Father, Son and Holy Spirit, and the extra He allowed the Holy Spirit to flow from both Father and Son. This is highly specialized word magic at its most fundamental level. (An analogy with physics is to see the Tetragrammaton as the most fundamental of elementary particles embodying the four forces – gravity, electromagnetism, and weak and strong interactions.)

The claim of natural magic that there exist absolute words is remarkable, and not readily comprehensible, to say the least. But nobody can doubt the existence of musical magic, the magic of melody, harmony, proportion and number. Words are practical everyday tools, so much used by all of us as aids to delineate thought and to communicate that it makes it difficult for us to see them as magical objects. But not so music. The fact that it has no such strict correspondence with things places it in a different category, which makes it easier to accept that music is itself a kind of magic. Is it not wonderful and mysterious that a flowing architecture of sound can be created that can act so powerfully on the mind? What material explanation of that can there be? Powerful musical forms exist through the creative genius of the composer and the empathetic performance of the musician or musicians. What is magic if not the manipulation of immaterial forms in this way? Music literally enchants. It is magic.

It is scarcely surprising that music appears as a magical force in the earliest myths. Orpheus, presented with a lyre by Apollo, who taught him its use, played such music that he enchanted the wild beasts and caused trees and even rocks to dance. It is reputed that a number of ancient mountain oaks at Zone in Thrace still stand in the dancing pattern where he left them. And in more recent times, Tolkein's creation text *The Silmarilion* takes music to be the primal stuff of the universe. Such also was the view of the Pythagoreans, whose insight into the mathematical and astronomical nature of music – the existence in harmony of simple numerical ratios, the music of the spheres – connected music with number and the stars in an intimate relationship. Musical magic and number magic were inextricably linked in Pythagorean thought.

Number magic, or numerology, appears strongly in the

cabbala. The hierarchy of God, angels and heavenly spheres is signified not only by integers, but also by letters of the alphabet. This meant that words could be arithmetized and that arithmetical operations could therefore be performed to produce ultimately the occult meaning of the word. Less elevated branches of numerology see all sorts of occult correspondences between things and numbers, none of which commands serious attention. But what is generally being affirmed by number magic is the existence of an occult, spiritual connection between things and number. Stripped down to its materialistic base, it is a connection that has been explored in great generality by mathematical physics, and there is no doubt that the results amply justify the basic intuition. But why mathematics works is still a great mystery. Its forms correspond to the motion of material objects. By manipulating the one we understand how to manipulate the other. Nowadays, musical magic has simply become music, and its offspring, number magic, has become simply mathematical physics. We tend too often to forget the magic which is still present.

The most down-to-earth branch of magic is elemental magic. In purporting to comprehend the nature of earth, air, fire and water it shades off at one extreme into recipes and crafts of all kinds, including witchcraft, and at the other it speaks in the overtly spiritual language of alchemy. Much of the substance of writings in this area consists of prescriptions of one sort and another – how to make poisons, how to treat ailments, and so on. But serious alchemy was the search for the magic ingredient, the Philosopher's Stone, that would turn base metals into gold or silver, and the search for the elixir of eternal life. No physician of the time was without some knowledge of applied chemistry. But, equally, he could not be indifferent to the Magical Theory of Everything – especially the doctrine of unity, of spiritual correspondence between metals and heavenly bodies, and of the connection between bodily organs and the stars as prescribed in the theory of microcosmos–macrocosmos. And he would be familiar with the great alchemical symbols of dragon (imperfect matter), marriage (Sun impregnating Moon to conceive the Stone), the egg (alchemical vessel hatching matter) and the tree (producing, as fruit, spirit from the earth). All of these summarize and adorn a subtle, esoteric branch of magic which underpinned the messy, smelly and often dangerous practice of chemical and medical craft.

When people think of magic it is usually elemental magic, with its images of witches' cauldrons and Prospero-like control of the forces of nature. When they think of magicians it is likely to be of those who claim striking powers, such as the ability to fly, to control the wind and the rain, to become invisible, to change form, to transport themselves instantly over large distances, to effect miraculous cures, to spell-bind enemies and hold them by enchantment. These powers were certainly worth pursuing, and no doubt many would-be magicians preferred the search for such powers to the more contemplative, intellectual activities implied by talismanic, word and musical magic. Those powers are still sought, but the activity is now what we think of as engineering and technology. We now know the conditions necessary for man to fly, what nerve gas will 'spell-bind', how to effect many cures. The control of wind and rain, invisibility, mutability and instant transport are currently beyond our powers and will, I suppose, remain so. Elemental magic on the one hand and applied chemistry, technology and engineering on the other share a common aim – the control of nature – but elemental magic did not work and science did. In this case magic had strayed into a realm in which it did not belong and was eventually evicted once empirical science got under way.

Another victim was the magical cosmology that underpinned everything. Beyond the illusion of multiplicity in the world lay a profound spiritual unity, the One of the Pythagoreans and the Neoplatonists. Everything was connected to everything else by hidden influences and correspondences which it was the task of the magi to elucidate. Based on Aristotle's teaching and the Pythagorean heavenly spheres, the cosmic picture was one of a vast spiritual hierarchy with the qualities of the higher spheres emanating downwards to the lower. This cosmology was comprehensibly elaborated by Robert Fludd (1574–1637), one of the last Hermetic philosophers, whose goal was to summarize the knowledge of the universe and man – macrocosm and microcosm. His universe was an immense spiral of 22 turns emanating from God and descending to the impure depths of the earth, each turn designated by an integer, a Hebrew letter and an essence. Number 1 is Mens, the Cosmic Mind, followed by the nine orders of angels – Seraphim, Cherubim, Dominations, Thrones, Powers, Principals, Virtues, Archangels, Angels. Below these come the stars, Saturn, Jupiter, Mars, the Sun, Venus, Mercury and the Moon, followed by the four elements, in order, fire, air, water and finally earth. In the doctrine of emanation lower beings are emanated as manifestations of the beings in the higher level. Thus, for example, Michael among the Archangels emanates as the Sun among the planets, as the heart in the body and as gold among metals. Actions taken at one level have repercussions at other ones – which is the basis of all magic. The idea of Plato's Great Year, in which the planets return to their original positions, finds its correspondence in the Fludd cosmology in the concept of an eternal cycle in which Sunset and the beginning of Night and Privation are followed by an Earth void and empty of divine power succeeded by Creation, Sunrise and an increase of vigour, then a decrease of vigour, then Sunset again. It was, indeed, a Theory of Everything – Everything Spiritual, that is. Magnificent achievement that it was, it could not survive the alternative vision of Kepler, Gilbert, Galileo and Newton, even though its subject-matter was spiritual rather than material.

Yet spiritual or natural, magic still survives, though it is not fashionable to say so. Only elemental magic has disappeared. People today still wear charms in the forms of rings, bracelets, pendants, etc. Christian young ladies wear gold or silver crosses; tough guys wear tattoos. What is this but talismanic magic? Poetry still literally enchants with its choice of words, and music still enthrals. Word magic, music magic, talismanic magic – in short, natural magic – being intrinsic to our world will always survive under one name or another, alongside religion and science.

But nowhere has magic exerted more power than in its manifestation as number magic in science itself.

Six

> Look how the floor of heaven
> Is thick inlaid with patines of bright gold;
> There's not the smallest orb which thou behold'st
> But in his motion like an angel sings,
> Still quiring to the young-ey'd cherubims:
> Such harmony is in immortal souls;
> But, whilst this muddy vesture of decay
> Doth grossly close it in, we cannot hear it.
>
> William Shakespeare, **The Merchant of Venice**, Act V (i)

Ever since Johannes Kepler discovered that planetary orbits were ellipses and Isaac Newton described the effects of gravity by a mathematical equation embodying his famous inverse-square law, mathematics has increasingly permeated physics, and in some areas it has gone a long way towards converting that empirical subject into a kind of mathematical theology. Mathematics is beautiful, mysterious and not without a touch of mysticism. Science without mathematical structures is now unimaginable, but this was not always the case. Mathematics was very much a part of magic before science got off the ground, and its magical influences can still be found today inspiring many theoretical physicists, as it did long ago for the followers of Pythagoras. The intimate relationship between mathematics and science that has developed over the centuries is now taken very much for granted. We are now no longer aware of the magic. This chapter and the next two explore this extraordinary symbiosis, how it began, what its limitations are and how our destiny is well and truly numbered.

According to one of the myths that grew up around the magical figure of Pythagoras, the music emitted by the planets and stars was heard by Pythagoras alone among mortals. The rest of us, the myth continued, cannot detect the harmony of the spheres, not because we do not possess fine enough senses, but because we have heard it from birth. The modern scientist will tend to write off all of that as merely a pretty metaphor, but it is not obvious that he is right to do so. From what we can gather from the various accounts of Pythagoras and his followers, they had a unique vision of the world that, duly seasoned with modern pragmatism and empirical censorship, is by no means inactive today, mystical though it may be at base. The interaction between outright mysticism and hard-headed science is less unusual than is generally supposed, and nowhere is it exemplified more vividly than in the modern belief that science can reveal the secrets of the whole universe and its creation.

Pythagoras is rather shadowy. He left no written works, and what discoveries may be attributed to him and what to members of the mystical brotherhood that he founded must be speculative. Born around 580 BC, he was raised in Samos in the Ionic colony of Greeks on and near the western shore of present-day Turkey. It is possible that he knew Thales and his school at Miletus, from which he may have been advised to visit Egypt, which he did, in order to absorb the ancient lore of that old civilization, and the mathematics passed down the ages from Babylon. At some stage a distinctive style of magical and religious teaching crystallized within him and attracted followers. Around 532 BC he and his disciples established themselves in Croton, a Dorian colony in southern Italy, and founded a mystical order devoted to Orphism, philosophy and mathematics.

This was an order that exemplified something deep in the human spirit which has never been lost. Their philosophy saw the sensible world as false and man's true life as being of the stars. They advocated a life devoted to theory in the Orphic sense of intoxicating contemplation, and they made its symbol the pentagram in the form of a five-pointed star. This vision of the Pythagoreans and their extraordinary revelations have excited and instructed men through 25 centuries. At the heart of their beliefs was the conviction that the real nature of things consisted in number.

The study of such topics was given the name mathematics by Pythagoras, who was in some sense the world's first theoretical physicist. His mathematics divided into two branches, one dealing with the discrete, the other with the continuous. The discrete was further divided into the absolute, meaning arithmetic, and relative, meaning music. The continuous was either static, meaning geometry, or moving, meaning astronomy. Arithmetic, music, geometry and astronomy became the four subjects of the medieval Quadrivium, which, along with the Trivium of grammar, dialectics and rhetoric, formed the basis for structuring knowledge throughout the Middle Ages in Europe.

Of the two branches of mathematics the Pythagoreans regarded the discrete as supreme. Above all was the number 1, its physical representation a point out of which all things were generated, its mystical representation, reason itself. Number 2 represented opinion, since the latter required ambiguity and, physically, it represented a line. Number 3 was the second triangular number (1 being the first), 6 the third, and 10 the fourth. Being the sum of the first four numbers whose ten points arranged in a triangle formed a revered symbol, the tetractys, 10 was the most sacred of all, representing perfection. A surface was associated with 3, since three points were needed to define it. Number 4 was associated with solids and justice (a square deal?) and 8 was a cube. Number 5, being the union of the first even and the first odd number, was connected with marriage, and the prime number was dedicated to the goddess Athene. The fundamental significance of the series of integers lay in its ability to be generated out of the One. Numbers were true, absolute entities, revealing timeless, eternal relationships that existed between them. Numbers were the basic reality. Apparently continuous things were really made up of units, ultimately the One.

This monism, this number theory of the world, was revealed most beautifully in the relative tonal qualities of music. A taut, vibrating string emits a note whose pitch is determined by its tension, its weight and its length. The Pythagoreans knew, and possibly discovered, that exactly halving the length raised the pitch by an octave, a musical interval regarded as fundamental by all peoples, whatever their culture. Here was a number becoming manifest in a clear, physical way – the number 2 connecting the tonic with its octave. Even today we speak of tuning the piano to 'philosophical pitch' when the frequency of middle C is 256 vibrations per second (256 Herz) because the the frequency of any C is then 2^n Hz, where n is an integer (equal to 8 for middle C).

Almost as basic to the ear as the octave is the fifth, for example, G in the key of C. The fifth is obtained by reducing the string to two-thirds of its original length. The next important note is the fourth (F), obtained by reducing the string to three-quarters of its original length. The interval between fourth and fifth is also fundamental, corresponding to a change in length by one-sixth, and it is called a tone. There are, therefore, six tones between tonic and octave, but more notes are needed in order to obtain the basic harmonies.

Harmonies are especially pleasing when simple numbers relate the frequencies. (No doubt there is an epigenetic explanation of this.) The diad consisting of tonic (C) and fifth (G) has a ratio 2:3, which is also the ratio for fourth (F) and octave (c). That for C and F is 3:4. These are truly fundamental harmonies as every human ear will affirm, and they are defined by very simple number ratios. Such a delightful example of the deep significance of number cannot fail to make an impression on the mind. Pythagoras must have seen this as number manifested as music and a powerful confirmation of his beliefs.

With such simple ratios emerging from diads it is natural to go beyond two-note harmonies and ask about three-note consonances. The simplest ratios 1:2:3 represent a basic triad consisting of tonic, octave and fifth above, which sounds too much like the simple diad 2:3 to be interesting. A similar remark applies to the next simplest triad, 2:3:4, because it again involves an octave. We get something interesting when we look at the ratios 3:4:5, because this involves a new note, the sixth. In the key of C this would correspond to CFA. If C is replaced by c, so the triad sounds the same, the notes FAc are in the ratios 4:5:6. These particular ratios actually define all the harmonic triads on the piano, realizable provided the notes D, E and B as well as A are added with the right frequencies. We then have the full scale of eight (hence octave) notes. The ratios are shown in Table 1.

This sequence of notes is the Lydian scale. To allow modulation from one key to another without all the notes having to be retuned, the purity of this scale must be sacrificed.

C	D	E	F	G	A	B	c
1	$\frac{9}{8}$	$\frac{5}{4}$	$\frac{4}{3}$	$\frac{3}{2}$	$\frac{5}{3}$	$\frac{15}{8}$	2

Table 1

Changing the pitch of the scale without changing the intervals requires the so-called equi-tempered scale in which there are now thirteen notes (sharps and flats added) separated by twelve intervals each equal to $2^{1/12}$. The comparison between the two scales in decimal notation is shown in Table 2.

Lydian	1.000	1.125	1.250	1.333	1.500	1.667	1.875	2.000
Equi-	1.000	1.123	1.260	1.335	1.498	1.682	1.888	2.000

Table 2

The use of such a scale, famous in the title of the J. S. Bach preludes, *The Well-tempered Clavier*, goes back perhaps to Aristoxenus (c. 350 BC), a pupil of Aristotle. The facility of modulation has to be paid for by having to put up with a slightly imperfect set of harmonic diads and triads. Only an excellent ear can detect the difference. Nevertheless, the equi-tempered scale is conceptually blasphemous in abandoning the sacred harmony of number revered by the Pythagoreans.

Harmony was explored in geometry as well as music. One of the Pythagoreans' greatest secrets was the discovery of the twelve-sided and twenty-sided 'spheres'. Of the five perfectly regular solids, the tetrahedron, the cube and the octahedron were almost certainly known to the Egyptians: Pythagoras discovered the regular twelve-sided solid, the dodecahedron and the regular twenty-sided solid, the icosahedron. It was, however, ironic that his famous law relating the squares on the sides of a right-angled triangle threw up the quantity $\sqrt{2}$, which cannot be expressed as a ratio of whole numbers. The ratio of the circumference of a circle to its diameter, π, was the same. This was a serious oddity. Rational numbers were the integers and those fractions that could be expressed as a ratio of integers. But the square root of 2 could not be expressed so, nor could pi. Such numbers were irrational, and so they are termed today. Their existence must have alarmed every right-thinking Pythagorean.

If rational number was music and geometry, must it not be manifest in the last of the Quadrivium, astronomy? Pythagoras probably knew that the Earth was spherical, perhaps because of the curved shadow on the Moon during an eclipse. Following curved mysterious paths around the Earth, as if inscribed on huge transparent spheres, were the Moon, the Sun, the five naked-eye planets (Mercury, Venus, Mars, Jupiter and Saturn) and the stars. The Pythagorean Philolaus (c. 500 BC) thought that the the planets, Sun, Moon and Earth moved around a central fire (the hearth of Zeus), whose light could not be seen directly because the Earth always faced away from it, but only indirectly from the Sun, and to some extent from the Moon, by reflection. Possibly to explain eclipses, and possibly to achieve for the astral world the perfection of the number Ten, Philolaus imagined that a dark Counter-Earth existed, invisible to the Earth.

From the Earth, however, the visible cosmos consisted of eight entities moving on their spheres. In order, from the Earth, they were deemed to follow the hierarchy: Moon, Venus, Sun, Mars, Jupiter, Saturn, stars. This hierarchy was the Babylonian one, which gave rise to the names of the days of the week. The first hour of Saturday was governed by Saturn, the second by Jupiter, and so on to the seventh hour governed by the Moon, when the sequence repeated. The twenty-second hour was again Saturn, the twenty-third Jupiter, the final hour Mars, and the first hour of the next day, the Sun, hence Sunday – and so on. The stars were appropriately above such a prescription. The same hierarchy described the seven ages of man. According to one scheme the Moon rules up to the age of 4, Mercury up to 14, Venus up to 22, the Sun up to 41, Mars up to 56, Jupiter up to 68 and Saturn up to 98.

Eight planetary beings meant that the facet of Number which manifested itself in the musical octave was manifesting itself in astronomy. Between the Earth and the stars the interval was an octave of six perfect tones. Between the Earth and Moon, one tone, a semi-tone between the Moon and Mercury, another semi-tone between Mercury and Venus, but a tone and a half between Venus and the Sun, so making the Earth–Sun interval a perfect fifth. A full tone between the Sun and Mars is followed by a semi-tone between Mars and Jupiter, a semi-tone between Jupiter and Saturn, and a final semi-tone between Saturn and the stars, making a perfect octave. This scheme suggests the scale shown in Table 3.

C	D	Eb	E	G	A	Bb	B	c
1	$\frac{9}{8}$	$\frac{6}{5}$	$\frac{5}{4}$	$\frac{3}{2}$	$\frac{5}{3}$	$\frac{9}{5}$	$\frac{15}{8}$	2

Table 3

Curiously, in this particular scheme (one favoured by Censorinus), the important note F is entirely missing. (Other schemes introduce an F.)[1] One consequence is that only seven harmonic triads are allowed. On this scheme the harmony of

Major	Planets	Minor	Planets
C E G	Earth, Venus, Sun	C Eb G	Earth, Mercury, Sun
Eb G Bb	Mercury, Sun, Jupiter	E G B	Venus, Sun, Saturn
D G B	Moon, Sun, Saturn	D G Bb	Moon, Sun, Jupiter
		C E A	Earth, Venus, Mars

Table 4

the spheres consists of the major and minor triads shown in Table 4.

Given the mythological characters commonly associated with the planets, what poetry and drama are suggested by these heavenly consonances!

For the Pythagoreans the harmony of the spheres was a reality. The slow cycles of the heavenly bodies were the measured bars of a vast, all-embracing symphony, audible to the enthusiastic mathematical ear. Harmony was everywhere as the highest good, and a man's health depended upon his achieving a relationship with the universe which partook of this harmony. This was best attained through the passionate involvement in mathematics which linked a man's soul to the eternal music of the spheres and opened his mind to the mystical significance of Number, above all, to the Number One manifest, not as attribute, but as essence, permeating the world.

Number also became associated with the planets via the magic square, familiar to the medieval Arabs. In a magic square sequential integers are so arranged that rows, columns and diagonals each separately add to the same number. The simplest is the square of 3 containing the first nine integers, assigned to the planet nearest the most refined regions, namely, Saturn. The squares of 4, 5, 6, etc. were sequentially assigned to Jupiter, Mars, the Sun, and so on to the Moon. The square of the Sun was particularly noteworthy in that rows, columns and diagonals added up to 111, and the sum of the six rows was 666 – 'the number of the Beast'. All rather sterile, but consistent with the idea that number was all embracing, though that claimed universality could easily give rise to numerological superstition, and did, and does so today.

Of course, long before Pythagoras, the Egyptians used and manipulated number for practical matters, much as most of us do today. To them numbers were tools to solve specific problems, and generalizations of any breadth did not occur, as far as we are aware, until Pythagoras. His passionate belief in the existence of the most intimate connection between the world and mathematics continued to inspire men down the ages, and no one more than Kepler 20 centuries on.

That strongly mystical strand of Pythagorean thought, indeed, lost nothing in its transmission by Plato and the Neoplatonists to the mind and soul of Johannes Kepler, mathematician and astronomer. Brought up on the Copernican theory of the solar system, which he embraced enthusiastically, Kepler found time, when he was not teaching mathematics in the Protestant seminary of Graz in Austria, to develop a remarkable theory to explain the number and size of the planetary orbits. This theory was published in a book entitled *Mysterium Cosmographicum* when the author was just 25 years old in 1596, and it was this book that brought Kepler's name to the attention of the Danish astronomer Tycho Brahe, and also of Galileo. The theory was pure Pythagoreanism.

In Kepler's time, as in previous times, only five planets – Mercury, Venus, Mars, Jupiter and Saturn – were known.

(Uranus was not discovered until 1781, the asteroids until 1801, Neptune until 1846 and Pluto until 1930.) After Copernicus, the Earth became a planet along with the others, orbiting around the Sun. The questions that intrigued Kepler were Why six and only six planets? And why did the orbits have the radii that were deduced from observation? At that time, like everyone else, he believed the orbits to be circular, and there were 23 years to go before his famous discovery that the orbits were, in fact, elliptical. But the problem that presented itself with all its mystery was not one connected with orbital shape; rather it was a question of ontology – Why six? Why those radii?

Kepler found an answer in a quantum principle connecting the old Pythagorean planetary spheres and solid geometry. In the latter there was the well-known quantization that he required: there were five intervals between the planets, and there are five, and five only, regular solid bodies. Here was the mystical explanation of the number of planets. As regards size of sphere in relation to the regular solids, Kepler explains in his book:

> The sphere of the Earth is the measure of all. Circumscribe about it a Dodecahedron: its circumscribed sphere will be Mars. Circumscribe a Tetrahedron about Mars: its circumscribed sphere will be Jupiter. Circumscribe a Cube about Jupiter: its circumscribed sphere will be Saturn. Now inscribe an Icosahedron within Earth: its inscribed sphere will be Venus. Inscribe an Octahedron in Venus: its inscribed sphere will be Mercury. Here you have the reason of the number of planets.

In this scheme Earth's sphere is wedged suitably between the discoveries of Pythagoras – the dodecahedron (the 'sphere' with 12 pentagons) and the icosahedron (the 'sphere' with 20 triangles).

In later years Kepler must have looked back on this youthful work with mixed feelings. Tackling the difficult problem of the orbit of Mars, as revealed by Tycho Brahe's patiently and accurately acquired data over many years, he came to insist on close agreement between theory and observation. Full of Pythagorean and Neoplatonic mysticism, and not averse to practising a little astrology, he nevertheless affirmed in his famous book – entitled what else but *Harmonices Mundi?* – that neither Mercury nor Mars, but Copernicus and Tycho Brahe were his stars. His approach to the problem of the orbit of Mars was what we might call today very professional, and it eventually led him to see without doubt that the orbit was not the perfect circle of Aristotelian astronomy, but a beautifully perfect ellipse. Moreover, he found that a clear and simple mathematical relation existed between the period of the orbit and the size of the ellipse. His brilliant speculation regarding the Pythagorean regular solids had to be abandoned. Nevertheless, here was number manifest for the first time as clearly as it was in music, and Kepler did not fail to elaborate his planetary laws in terms of musical harmonies. He believed that he had surpassed the Pythagoreans and had discovered the true music of the spheres, and in a quite definite sense, he had.

The power of Pythagorean discoveries remains undiminished. The versatility of the element carbon to combine with hydrogen in myriad ways was exploited some years ago by a group of chemists in the true Greek spirit. They synthesized a hydrocarbon molecule consisting of 20 carbons and 20 hydrogens in the shape of a dodecahedron – one of the Pythagorean regular solids. They said that they were motivated to produce a dodecahedron by its 'exquisite shape' and 'especially high aesthetic appeal'. Dodecahedrane joins cubane and tetrahedrane, and perhaps by now or soon there will be octahedrane and icosahedrane.

The mystical power of the Pythagorean intuition, still very much alive and well, was the trigger that shot mathematics into the physical world. But, just as Hermetic fervour was paralleled by a cooler look at nature, the cool look that allowed science to develop, so cooler currents were there to expand mathematical thought. And that expansion is truly magnificent. But, just as science has strayed into areas where its usefulness is questionable, so mathematics has invaded practically every corner of our lives, mostly benevolently, but not always. Just as we need to know what the nature of science is and what are its limitations, so we need to know what mathematics is and what are its limitations.

Seven

He that but once too nearly hears
The music of forefended spheres,
Is thenceforth lonely.

Coventry Patmore, 'The Music of
Forefended Spheres'

Pythagoras gave us the vision of the world as Number, Plato of a world as transcendental forms; Aristotle bequeathed the laws of logic, Euclid deduced the properties of space from a set of axioms, Archimedes applied mathematics to physics and Diophantus gave us the first work on algebra. The Greek legacy is staggering. It provided a base that was to prove invaluable in the rational study of nature and of mathematics itself. The mathematical interpretation of the world became the aim of physics, even at the risk, and the risk is great today, of outstripping the empirical base. It is easy to see, even from the original Greek impulse, how this has come about. It may be noted that the reductionist idea of all objective knowledge being reducible to the laws of physics really means reducible to mathematical structures.

After Kepler there could be no science without number. Galileo was convinced that the book of nature was written in the language of mathematics, and Descartes, though distinguishing between physical and purely mathematical explanation, insisted that the subject-matter of science ought to be restricted to what could be expressed in mathematical form. Some modern scientists, while agreeing with Descartes, go further and adopt the view that what cannot be expressed in mathematical form is meaningless. Descartes, who was the first to distinguish mind and matter as irreconcilable kinds, would have regarded this as nonsense.

Descartes's important contribution to mathematics was to apply algebra to geometry. As all of us who remember Euclidean geometry from our schooldays will know, the proofs of the various theorems used ordinary language and this did not always make them easy to follow. Anything out of the ordinary seemed to require exceptional ingenuity in manipulating the basic axioms. The Cartesian invention of coordinate geometry made many proofs, notably those that applied to conic sections like the ellipse and the hyperbola, much easier to demonstrate, and it allowed others to be discovered.

He made another important contribution to science in distinguishing *mathematical* explanations from *physical* explanations. In spite of his insistence on the mathematical he did not, in fact, rule out physical explanation. For him and for many others the motion of bodies could come about only through physical contact. Thus, a physical explanation of the motion of the Earth around the Sun, the only intelligible one, was that the Earth was carried by the motion of the fluid in which it was immersed. Descartes's Theory of Vortices applied to all the planetary motions. Implicit was the dismissal of the idea of action at a distance as magical nonsense.

This view is, of course, supported by our common experience – pushes and pulls had to involve things in contact. In his experiments in mechanics, Galileo dealt with things in contact, and gave mathematical expression to these contact laws of motion. It was Newton who opened a gulf between physical and mathematical explanation, a gulf that has haunted science ever since. Common sense, based on common experience, knew that to keep things moving you had to keep pushing them, and that any pushing or pulling could be done only if you were in contact with the object being pushed or pulled. Newton confounded common sense through the conceptions of inertia and action at a distance. All bodies possessed inertia, which meant that their motion would not change unless they experienced a force, but that force need not be a contact force. He discovered a simple mathematical formula that could mathematically explain all gravitational phenomena but could not physically explain anything. How the Moon was kept in orbit around the Earth, or a planet around the Sun, was due to the force of gravity, a magical action at a distance, and the magical property of bodily inertia. Common sense has never recovered its authority in this field. Applied to the observations made of motion on Earth and of motion in the heavens, Newton's mathematics works. The universal inverse-square-law of gravitation could explain Kepler's laws of planetary motion, and Newton's laws of motion could explain all of Galileo's results; even the ocean tides were explained. But how action at a distance works *physically* was *not* explained.

The publication in 1687 of Newton's *Philosophiae Naturalis Principia Mathematica* was an event of unmatched magnitude. In the preface, Newton writes:

> and therefore I offer this work as the mathematical principles of philosophy, for the whole burden of philosophy seems to consist in this – from the phenomena of motions to investigate the forces of nature, and then from these forces to demonstrate other phenomena.

Philosophy here means, of course, natural philosophy, i.e. science. This statement defines what science is about, even today. But, as Newton goes on elsewhere in the *Principia* to say:

> The main business of natural philosophy is to argue from phenomena without feigning hypotheses, and to deduce causes from effects, till we come to the very first, which is certainly not mechanical.

Thus, physical explanations are to be avoided and only mathematical relations admitted – at least at first. Newton himself thought that gravity as a force acting at a distance was an absurd concept, but his belief in God who had created a world understandable through mathematics allowed him to think that the source of gravity might be immaterial.

This disposition of Newton to imagine the possibility of an immaterial force is an example of certain mystical tendencies in his philosophy. He was well aware of Rosicrucianism, and his unpublished manuscripts show his interest in alchemy. A Pythagorean streak exhibits itself strongly in his theory of colour mixing, in which seven rainbow colours are depicted by arcs of a circle proportional to the seven musical intervals of the octave – five tones, green, blue, violet and red, and two semi-tones, orange and yellow. Not only shades of the music of the spheres here, but rainbow-coloured ones.

After Newton the belief in God the Mathematician prevailed. Sceptical attitudes, such as the earlier exhortation of Bacon to stick to experiment –

> It cannot be that axioms established by argument can suffice for the discovery of new works, since the subtlety of nature is greater many times over the subtlety of argument[1] –

were ignored. That mathematics could explain the world was

put down by Leibniz to the existence of a pre-established harmony between the two. Mathematical insight was all important, even superior to logic, according to Descartes:

> Intuition is the undoubting conception of a pure and attentive mind, which arises from the light of reason alone, and is more certain than deduction.[2]

A similar sentiment is shown by Pascal:

> Our knowledge of the first principles, such as space, time, motion, number, is as certain as any knowledge we obtain by reasoning.[3]

For both of these mathematicians, mathematics was an intuitive affair, and for Pascal, a devout Catholic, science was a kind of worship. The world of God was to be known through mathematics.

The expansion of mathematics was spectacular. The differential calculus developed by Newton and Leibniz was used by Lagrange to treat mechanics entirely mathematically without any reference to actual physical processes. He saw that one of the ways in which nature behaved could be summarized in the Principle of Least Action, which lay at the heart of a new mathematics – the calculus of variations. An example is that of light, which travels along a path that takes least time, i.e. a straight line in a uniform medium. Another example is the straight line motion of a body unaffected by forces. This principle, which turned out to be very powerful, was used by Lagrange, and later, after some generalization, by Hamilton, to produce the most widely applicable purely mathematical formulation of classical mechanics. Action is a mathematical creation, defined by the quantity produced by multiplying energy and time, or momentum and distance.

The principle states that a mechanical system will follow an evolution such that its action is a minimum. Keeping the action as small as possible was seen by some, perhaps, as a kind of thrifty behaviour of nature. Nowadays nature is no longer endowed with human qualities, so nobody thinks like that any more – there is just this mathematical product that keeps small, that is all. But action, a purely mathematical creation of classical mechanics, becomes very real indeed in quantum mechanics, though still, or even more, incomprehensible.

Another glittering achievement of classical mathematical physics was the creation of the electromagnetic field and the synthesis of all electrical, magnetic and optical phenomena in one theory, by Maxwell. The individual laws of current electricity and electrostatics discovered by Ampère and Coulomb, the laws of magnetism by Gilbert and Øersted, the laws of induction by Faraday, were all accounted for, provided that, once again, we accept the electrical and magnetic forces as acting at a distance. That light travelled at a finite speed was discovered in 1676 by Römer. Its speed, measured in 1849 by Fizeau in the laboratory and more accurately a year later by Foucault, was also a facet of the theory. All these strands were embodied in four coupled differential equations which Maxwell published in 1873. Classical physics now had two so-called fields, that of gravity and that of electromagnetism, both deeply mysterious elements of nature that featured this magical action at a distance, a mystery that could not be dispelled by mathematics. It seemed that the mathematization of physics was complete. But more mathematics was to come.

If irrational numbers like $\sqrt{2}$ and π made the Pythagoreans blanch, how would they have responded to $\sqrt{-1}$, the square root of minus one, which does not exist? As a symbol it found its use in connection with vectors. The concept of a vector – a quantity, like velocity, with both magnitude and direction, as distinct from a scalar, like speed, with magnitude only – was a familiar one in the seventeenth century. Using the imaginary quantity $\sqrt{-1}$, usually denoted by the symbol i, it was possible to develop a vector algebra in which a vector could be represented by a single number, albeit complex, that is, consisting of a real plus an imaginary part, for example, $A = B + iC$. Thus, B could quantify the component along the x-direction and C could quantify the component along a direction at right angles, say, the y-direction. Resolving the vector into two mutually perpendicular components meant that magnitude and direction could be summarized by one complex number. (A lot of mathematics is summarizing.) Differential vector algebra followed, and a whole library of functions of a complex variable developed. Such a library is an essential possession for any physicist applying theory to the prediction or interpretation of experimental results.

Given that a Pythagorean was robust enough to survive the shock of imaginary numbers, would he survive the shock of non-commutability? We are all familiar with the arithmetical equation $7 \times 6 = 42$ and we would be rightly stupefied by somebody claiming that although $7 \times 6 = 42$ is true, 6×7 is something different. In ordinary arithmetic the order of multiplying one number by another is irrelevant to the resulting product. Quantities that behave like that are said to be commutable. The product of non-commutable quantities a and b depends on the order of multiplication: $a \times b \neq b \times a$. We live in a four-dimensional world, one dimension of time and three of space. Suppose we wish to summarize the properties of a four-dimensional vector **A** using imaginary numbers. We can write $A = B + iC + jD + kE$, where $i = j = k = \sqrt{-1}$. It turns out that if we wish the *magnitude* of A to be given by a real number, so that we maintain contact with reality, then the product of i and j is not the same as the product of j and i. In other words, the imaginary numbers i and j do not commute as ordinary numbers do. These quantities are called quaternions, and they have the property $ij = -ji$, $jk = -kj$, $ki = -ik$. These are the rules for manipulating products of different sorts of imaginary number. We have a new mathematics.

Another shock to the inherited Greek sensibility came when after much endeavour it was realized that Euclidean geometry could not be proved to be true. In particular, the axiom asserting that parallel straight lines never crossed was impossible to prove. This axiom then had to be seen as defining a particular geometry, namely, that of Euclid. This meant that the Euclidean conception of flat space used by physics and apparently self-evident was not provably the case. The intuition of Euclidean space claimed by many, including Pascal and Kant, to be a truth was shown to be undecidable. This opened the way for the development of non-Euclidean geometry by Gauss, Lobachevsky, Bolyai and Riemann. One could imagine a spherical geometry in which parallel lines actually met, or a hyperbolic geometry in which parallel lines diverged. In spherical geometry the angles of a triangle add up to more than 180°, as in fact do the angles of a triangle drawn on the earth's surface. In hyperbolic geometry the angles add up to less than 180°, as is the case for triangles drawn on a saddle. Here was a new branch of mathematics that seemed to exist in its own right, with nothing to do with the real world.

The dethroning of Euclidean geometry opened people's

eyes not only to other geometries but also to empirical considerations. If space itself cannot be taken as given, how do we find out what it is really like? We know to an excellent approximation that Euclidean geometry holds locally on Earth, but what about the rest of the universe? How do we tell? We know that, locally, space is flat by seeing objects around us having location and relative position and performing measurements, such as adding up the angles of a triangle and seeing how close we get to the Euclidean prediction. The operative word there is 'see'. To find out geometry we must use light. Following concerns about the effect of the Earth's motion on the velocity of light and the experimental evidence from the famous Michelson–Morley experiment that there was no effect whatsoever, Einstein developed his Special Theory of Relativity. We obtain information about the world using light, and the structure of the world that we derive is determined by its properties. That structure was determined by Einstein to be that obtained when the velocity of light in vacuum is taken to be a fundamental constant of nature. Time was measured by a local clock and space was measured by sending out light pulses and receiving their reflections. In a penetrating analysis of the operational meaning of simultaneity, Einstein showed that there was no universal meaning to be attached to space on its own and time on its own but a universal meaning could only be attached to the four-dimensional continuum of space-time. Observers travelling at different uniform velocities, using the same clocks and using light in the same way, see space-time carved up differently into a particular 3D space and a particular time sequence. Since nothing travels faster than light, no paradoxes appear of the sort celebrated in the limerick:

> There was a young lady called Bright
> Who could travel much faster than light.
> She set out one day
> In a relative way,
> And returned on the previous night.

But we have to accept the fact that the passage of time for, say, a fast meson is much slower than for a stationary meson – a fast meson is observed to live longer.

Another far-reaching consequence of special relativity is embodied in the famous equation $E = mc^2$, where E is the total energy, m is the mass and c is the velocity of light. This equates energy and inertial mass. Conceptually, the mass of a body that determines inertia and appears in Newton's Laws of Motion is different from the mass that responds to gravity, yet it is found experimentally that these two masses appear to be identical. The Principle of Equivalence asserts that these two masses are actually identical, and not just approximately equal to one another. If this is the case, it should not be possible to distinguish the effect of a uniform gravitational field from an equivalent acceleration. The further equivalence of mass and energy means that light should be affected by a gravitational field. In flat space this would mean that the velocity of light would change. It follows that if the velocity of light is a universal constant, which it must be if we are to map out space and time for ourselves, then space in a gravitational field cannot be flat. Einstein, in his General Theory of Relativity, concluded that the idea of gravity as a force acting at a distance could be abandoned, and that gravity was really the effect of matter-energy on space-time itself. In general, space-time was a non-Euclidean four-dimensional continuum.

With Einstein's General Theory of Relativity we reach

the end of classical physics. The mechanics of bodies is described by the general equations of Lagrange and Hamilton; the phenomena of electromagnetism are deducible from Maxwell's equations. Both schemes can be made compatible with special relativity in the equations of electrodynamics. And the space-time in which events occur is described by Einstein's General Theory. All the equations are differential equations and, of course, boundary conditions are necessary in order to define solutions. It is by no means always possible to define these with precision and consequently solutions that are obtained may have only statistical significance. In the case of complex systems, concepts relating to chance and randomness inevitably appear. Furthermore, describing conditions at the boundaries requires a specification of the environment. If a microscopic description of a many-body system is out of the question, it is nevertheless possible to obtain general properties from the laws of thermodynamics or its microscopic counterpart, statistical mechanics. It is only here that the arrow of time that we all experience becomes evident. In all the equations of classical physics involving time, time is a coordinate like space and and can be positive or negative without affecting the basic physics. Only in the realm of large numbers and systems that are not isolated from the rest of the universe does time have a more realistic meaning.

When the equations of classical physics are applied to the whole universe, the assignment of boundary conditions becomes an uneasy affair. Cosmology, wishing to describe the evolution of the cosmos, must worry about defining a universal time, and be concerned with the physical conditions at $t = 0$. The universe is expanding. If it has been expanding since $t = 0$, then at $t = 0$, the Big Bang, the physical conditions will be far outside our experience. What one does about that is an interesting point to be taken up later.

Planck's discovery of the tiny quantum of action, symbol h, and the subsequent apprehension of the non-classical behaviour of electrons and light meant that any physical, as distinct from mathematical, description of the world was very difficult to imagine, and only through an abstract mathematical theory could a sort of understanding be obtained. At least in classical physics one could point to planets, molecules, or atoms that the mathematical description was about. A physical picture went along with the mathematical one. An electric current was a flow of little particles called electrons through a matrix of ions. Light was a wave, but of what was always unclear. With the discovery of quantum behaviour electrons were found to behave like waves and waves like particles, depending on what experiment was being carried out. A fundamental unpredictability concerning some of the properties of these quantum particles forces us to use the concepts of probability at a basic level. Nowhere is science more reliant on mathematics than here. Moreover, the intrinsic non-locality of quantum phenomena that is observed in experiment appears to force a mathematical description that, in the limit, must encompass the whole universe.

If the unsolved problem here is how to tie quantum behaviour in with gravitation, the success of quantum electrodynamics is extraordinarily impressive. Quantization of the classical equations of electrodynamics was comparatively straightforward, but dealing with its consequences for electrons, and for the electromagnetic field itself, revealed a world far removed from the classical picture and even further removed from the common-sense world we are familiar with. The electron was always considered as a point particle, but this meant that its self-energy, associated with the Coulomb interaction of one part of the electron on the rest, for example, was, in classical physics, infinite. In the full quantum electrodynamic theory other sorts of energy enter, and to complicate matters, the electromagnetic field surrounding the electron was extremely active, bubbling with particles continually coming into being and disappearing, the so-called virtual particles. Nevertheless, by a wonderful sleight of hand called renormalization, quantum electrodynamic shifts in the energy of the electron in a hydrogen atom could be calculated to an impressive precision and, more to the point, these calculations agreed with experiment. But the world was for ever changed. The vacuum was no longer featureless and empty. Instead, it bubbled with virtual excitations – virtual, because the excitations had only a transient existence – and this bubbling consisted of the appearance and disappearance of particles and anti-particles, the energy of formation being allowed by the Uncertainty Principle provided the duration of the particles was short. But a bubbling vacuum meant that it had energy and therefore, through the famous equation $E = mc^2$, it had mass. A vacuum with mass is somewhat worrisome for any future theory of quantum gravity.

In this far-flung realm of experience we depend on mathematics as never before; physics could never have got on without it. But is mathematics as reliable as we think it is? A common view, generated from our school experience of Euclid, is that mathematics deduces truths from a set of axioms deemed true. I would guess that no modern mathematician would go along with that.[4] In his book *Mysticism and Logic* Bertrand Russell said:

> Mathematics may be defined as the subject in which we never
> know what we are talking about nor whether what we are
> saying is true.

Mathematics, in other words, is a structure, logically self-consistent, it is hoped, in which case it can claim a kind of truth. Whether it is of use to science is another matter; empirical truth is quite different from mathematical truth. G. H. Hardy is famous for his toast:

> Here's to pure mathematics! May it never have any use.

Mathematicians are divided as to what mathematics is about, as regards both the question of its truth and questions concerning its nature. Descartes and Pascal would claim intuition of truths to be the source, and they would be supported by the philosopher Kant. Others would claim that mathematical structures, like Plato's forms, existed out there in their own right, waiting to be discovered. Gödel maintained that all sets were real objects:

> It seems to me that the assumptions of such objects is quite
> as legitimate as the assumption of physical objects and there
> is quite as much reason to believe in their existence.

There are questions of epistemology here – what is the source of mathematical truth? and questions of ontology – what mathematical truths exist in the world waiting to be discovered?

A quite different view is that mathematics is an invention of the human mind and nothing to do with truth. Russell and Whitehead's attempt to reduce mathematics to pure logic failed, but Hilbert saw mathematics as a purely formal mental structure that defined axioms, including those of logic. In principle, Hilbertian mathematics has nothing to do with nature at all, and Russell's definition of mathematics, as the subject in which we never know what we are talking about, applies a fortiori. Another approach, associated with the names of Zermelo and Fraenkel, which stemmed from the Russell–Whitehead work, was to explain mathematics in terms of set theory. Thus, the number 2 is the set of all systems containing two members, all swans that are white belong to the set of white swans, and so on. Of set theory, Hausdorff complained that it was:

> a field in which nothing is self-evident, whose true statements are often paradoxical, and whose plausible ones are false.

Russell was keen on set theory until he thought of a horrible paradox. He could distinguish two sorts of set, R sets and Non-R sets. An R set is one that contains itself as a member. A library catalogue, the set of books in the library, is itself a book and is therefore part of an R set. The set of mathematicians is obviously not a mathematician, so this set is a Non-R set. So far so good. Now consider the set containing all Non-R sets. Is this set R or Non-R? If it is R, it is a member of the Non-R set, which it is not. If it is Non-R, it is not a member of Non-R sets, but it is. The set of all Non-R sets cannot be described as R or Non-R, so what sort of animal is it? Set theory has its problems.

But worse was to come. Gödel makes the unkindest cut of all, proving that any mathematical system that is extensive enough to embrace the arithmetic of whole numbers is either incomplete or inconsistent. What Gödel did was to establish an arithmetic formula, A, with the meaning 'the truth or falsehood of A cannot be determined'. If A is true, arithmetic is incomplete – there is no axiom from which it can be deduced that A is true. Conversely, if A is false, then the arithmetic used to construct A is inconsistent. Consistency, as Morris Kline states in his book *Mathematics*, cannot be established by the logical principles adopted by the several foundational schools, the logicists, the formalists and the set theorists. But even proofs of any sort have been regarded with a degree of scepticism by at least some philosophers. Kline quotes Nietzsche:

> The virtue of a logical proof is not that it compels belief but that it suggests doubts

and Popper:

> There are three levels of understanding of a proof. The lowest is the pleasant feeling of having grasped the argument; the second is the ability to repeat it; and the third or top level is that of being able to refute it.

Mathematics has been a boon to science, in spite of all this. There is always the necessity of agreeing with what is observed that can act as arbiter, but even this test is never as simple as it seems. Observations, we are warned, are always theory-laden, so a test of a particular theory is never as clear as one would like. Kline quotes the concern voiced by Courant:

> A serious threat to the very life of science is implied in the assertion that mathematics is nothing but a system of conclusions drawn from the definitions and postulates that must be consistent but otherwise might be created by the free will of mathematicians.

Whatever mathematics is, it is not like the simple picture we have of it from our schooldays.

The exploration of the unfamiliar realms of nature, particularly the quantum and high-energy regimes, is yet empirical, as all scientific explorations must be. There exist raw data that must be interpreted to create an understanding, and here mathematics enters to provide a structure on which to base an understanding. But mathematics is glorious in the number of available structures; there may be many choices available, especially in the more esoteric fields of research. In these cases what criteria are we to use? Elegance and thrift, certainly, but given a number of elegant and thrifty theories, what then? Are we to have our understanding determined by the mathematician or research group that shouts loudest? Or may we hold on to a belief that reality itself will somehow continue to guide our choice?

In the latter half of the twentieth century, the mathematicians/scientists, having solved all simple problems, or having found that some simple problems are really very deep, have turned, with a certain amount of hubris, to the study of complex systems and, in so doing, have discovered some remarkable things. I started my book *Time, Space and Things* some thirty years ago with the sentence, 'Physics is about the simple things of the Universe.' I have not felt that it ever needed changing, in spite of the incredulity voiced by non-physicists. It is precisely by choosing 'the simple things of the Universe' that physics has been so successful. But that choice has meant leaving out the bulk of the real world; it has meant leaving out life itself with its complex structure and its interactions with the environment; it has meant leaving out all those human interactions that are central to our lives. But that is no longer the case. Enter the study of complexity.

Easing out from simple billiard-ball behaviour, the would-be student of complexity is bound to hit the topic of non-linear dynamics, a topic that is actually quite central to the study of a number of straightforward physical systems. Some of these are mechanical (such as the weather), electrical (such as the semiconductor that exhibits that wild departure from Ohm's Law known as a negative differential resistance), or magnetohydrodynamic (such as those plasma instabilities that frustrate the achievement of nuclear fusion and the goal of unlimited energy). Non-linear dynamics paints a picture of an unstable world in which a minute disturbance can evolve into a formidable entity, like a tornado in the case of weather, a high-field travelling domain in the case of one sort of a negative differential resistance or into an electric spark in another sort, or into wild wave-like oscillations in a plasma. In many cases those systems evolve into chaos. But chaos is no longer what it used to be – lawless and formless. To some degree it has been tamed, chaotic behaviour has been classified, and a language to talk about chaos has evolved. Impressive though this is, there are two even more striking results of this comparatively new science. The first is that, although all of this non-linear dynamics is sited firmly in the field of classical physics, with effect following cause in a determinate way, there is sometimes no way of predicting how the system will evolve. In those systems where this is true an infinitesimal change in the starting conditions produces an entirely different evolution. What price determinism in a system whose evolution cannot be predicted! The second striking thing is that out of complexity can come simplicity. Out of incredibly complicated interactions with matter individual forms can emerge. Such emergent forms are tornadoes and electric sparks, and no doubt life itself.

There are few smooth surfaces and few simple forms in nature. Think of a tree. It begins simply enough with a trunk, but then the trunk branches, and the branches themselves branch and produce twigs that produce further twigs. Shift to your own body. Think of your aorta branching into two large arteries which themselves branch and branch again to form the myriad pattern of blood vessels throughout your body. Think of a major river dividing similarly into multiple streams in its delta. On a different tack, think of a coastline with its innumerable indentations. None of these is exactly a simple geometrical pattern. Yet it turns out that they can be quantitatively comprehended by thinking mathematically in terms of a fractional dimension. The mathematician Mandelbrot, who pioneered the study of complex patterns, called these patterns fractals. The tree trunk, branches and twigs, looked at in a simple way, are all one-dimensional, yet the whole tree fills up three-dimensional space. The same with blood vessels. Both are fractal. A coastline looks like a smooth one-dimensional curve from far above the Earth, but, close up, the indentations occupy a two-dimensional surface. A coastline is also fractal. And the indentations have indentations. The property of a complex form in nature to repeat itself at smaller and smaller scales is very common. It was once thought that blood vessels in the lungs multiplied exponentially as they branched, but their pattern is better described as being fractal.

Encouraged by its success in its studies of chaotic systems and chaotic-looking shapes, the science of complexity is ready to tackle the problems of life – life's origins, its evolution and the myriad facets of human culture. James Gleik, in his book *Chaos*, mentions the extraordinary similarity of mathematical pattern between the distribution of large and small earthquakes and the distribution of personal incomes in a free-market economy. It is inevitable, and indeed sensible, that the precise language and thinking that derive from the physical sciences and from the mathematical insights into chaos and fractals will be used to guide the description of human society, providing analogies and suggesting parallels. But the operation has to be done with both sense and sensibility. Earthquakes and personal incomes may exhibit the same pattern. So do, for example, all face-centred cubic crystals. But you would not persuade information technology to give up silicon for germanium, nor would you expect the free-market economy to fund research into earthquakes on the strength of a similar pattern.

Earlier, I referred to the mathematician/scientist. Such a hybrid is commonplace in the physical sciences, but it is not obvious that such a hybrid can exist in the study of human society. Mathematics, maybe; science, impossible.

The studies of economics and sociology have long been invaded by mathematics, with effects that many regard as pernicious except, perhaps, in the application of statistics to simple measurables. However, a transformation of these studies into a natural science would require the existence of millions of similar societies in order to allow comparative investigations to be made. Such an ensemble does not exist, and among those societies that do exist, extensive experimentation is not feasible. In such circumstances mathematics can have only a limited part to play, and to forget that fact is to stray into scientism. But quantification is now universal. Obeying the commandment not to call meaningful what is not quantifiable, scientismic management seeks to benefit humanity by eliminating the need for common sense. Values, once thought best left in the hands of professionals, become redefined and given a number and a name. Quality can be measured. Statistical norms and deviations provide the criteria for assessment, doing away with the awkward and potentially devisive necessity of recognizing individual talent. Mathematics, of a sort, is rife. The usual justification is to provide an objective means of reaching decisions on the distribution of scarce resources. Suggesting that there are common-sense means of doing the same thing without using those scarce resources to institute largely meaningless measurements that disrupt and change the character of what is being measured seems to be, in the present quasi-religious climate, rank heresy. It might help if there were more awareness of the elementary empirical criterion of reducing to a minimum the disturbance by the act of measurement to what is being measured. Where this elementary criterion is not applied it is even more difficult than usual to attach any useful meaning to any results so obtained.

No doubt the driving ideology is that all things must become like physics, with the objectivity and qualitative precision afforded by mathematics. But human life is not like that. Nor can it be remotely imagined that entities like the great dimensionless numbers of physics will be found that define society in a fundamental sense. These numbers, formed from the fundamental constants of nature, inexplicable and mysterious, are truly magical in their effect on the imagination.

Eight

I believe there are 15,747,724,136,275,002,577,605,653,961,181,555,
468,044,717,914,527,116,709,366,231,425,076,185,631,031,296 protons
in the universe, and the same number of electrons.

Sir Arthur Eddington, **The Philosophy of Physical Science** (1938)

The size of each of the fundamental constants of physics is, of course, determined by nature and by the standard units of measurement – the metre, the kilogram and the second – which are defined by international agreement. Thus, the velocity of light in a vacuum (symbol c) is $2.997,924,58 \times 10^{8}$ metres per second (ms^{-1}), Planck's constant (symbol h) is $6.626,075,540 \times 10^{-34}$ Joule-second (Js), the permittivity of the vacuum (symbol ε_0) is $8.854,187,817 \times 10^{-12}$ Farad per metre (Fm^{-1}), the gravitational constant (symbol G) is $6.672,598,5 \times 10^{-11}$ cubic metre per kilogram per second squared ($m^3kg^{-1}s^{-2}$), and so on. These numbers would all change should we decide to express them in different units, say in the system yard, ton, fortnight (thought, no doubt unfairly, to be the natural set of units for describing the movement of freight on British railways, typically taking a fortnight to shift one ton one yard). Nature would not change – we would all feel as heavy as we did before whether we measured weight in tons or kilograms. The actual magnitudes of the fundamental constants have specific dimensions that have been decided by committee. If there are significant numbers that quantify something fundamental about nature, it is clear that they must be those that are dimensionless. Just as Pythagoras discovered long ago that it was not pitch itself that was important but the ratio of pitches – the factor of 2, for instance, relating the tonic to its octave – we find that there exists a fundamental set of dimensionless numbers formed by taking certain ratios of combinations of fundamental constants.

The existence of pure numbers in nature has suggested to some that their magnitudes should be intuitively predictable. Nobody was more convinced of that than Sir Arthur Eddington. In his 'Fundamental Theory' he pursued the principle that these numbers could be logically deduced from a few simple qualitative assertions about the world.[1] In deducing an equation to give the observed ratio of the proton and electron masses he points out that in relativity the ordinary momentum vector, the quantity obtained by multiplying mass by velocity, has 4 components and the energy tensor has 10 components, but if spin angular momentum is included the number of components increases to 10 and 136 respectively. He then considers a standard particle of mass m_0 which has an external energy corresponding, at low velocities, to its rest-mass, and an internal energy corresponding to internal motion. Considering now the hydrogen atom, he splits the total motion into the motion of the centre-of-gravity associated with the total mass $M = m_e + m_p$, where m_e, m_p are the electron and proton masses, and the relative motion of the electron with respect to the proton associated with the reduced mass $\mu = m_e m_p / (m_e + m_p)$, which is standard procedure. However, as far as the internal energy is concerned the mass m_0 is associated with 136 components of the energy tensor, and so $\mu = m_0 / 136$, but for the external motion there are only the usual 10 components, and therefore

$M = 136m_0/10$. Consequently, the masses of the electron and proton can be obtained from the quadratic equation:

$$10m^2 - 136mm_0 + m_0^2 = 0,$$

and so the dimensionless ratio $m_p/m_e = 1847.6$ is obtained. This has to be compared with the measured value of 1836.2. The comparison seemed good enough in 1931 when Eddington published his calculation.

Eddington's principle is a very exciting one, but there are few nowadays who believe his approach to be well founded. 'Fundamental Theory' was extraordinarily ingenious in deducing a simple equation that gave the proton-electron mass ratio. Another success was in predicting the magnitude of the reciprocal of the fine-structure constant, a dimensionless number, familiar in spectroscopy, a measure of the strength of interaction between the electromagnetic field and the electron. The reciprocal of the fine-structure constant is very nearly a whole number, namely, 137, which is enough to arouse the ambitions of any red-blooded Pythagorean.

Another spectacular prediction of Eddington's was the number of particles in the universe, as the quotation at the beginning of this chapter testifies. The cosmical number is $N = (\frac{3}{2}).136.2^{256}$, that is, one-and-a-half times the number of components of the full energy tensor times 256 octaves, which is the number quoted if the factor $(\frac{3}{2})$ is ignored. It is worth while following a simple argument for assessing the magnitude of N. The idea is that if R is the radius of the universe and N is the number of hydrogen atoms, the uncertainty in spatial position has to be $\Delta x \approx R/N^{\frac{1}{2}}$. This can be equated to the classical electron radius, $e^2/4\pi\varepsilon_0 m_e c^2$, where e is the charge on the electron, ε_0 is the permittivity of the vacuum and c is the velocity of light in vacuum. Now the radius of the universe is related to the rate of expansion of the universe as measured by the Hubble constant H according to $R = c/H$. According to general relativity, if the universe is a spherical, closed universe, the radius of the universe depends upon the total mass of the universe M. Now, if one looks at the spectrum of light emitted by stars and compares features in the spectrum with known spectral features of the elements observed in the laboratory, one concludes that by far the most common element in the universe is the simplest of all, hydrogen. Thus, to a good approximation, the total mass of the universe is that of all the hydrogen, i.e. Nm_H, where m_H is the mass of a hydrogen atom. The fundamental relation that connects the radius of the universe with its mass is $2GM/c^2 = R$, where G is the gravitational constant. Putting all of this together one gets $N^{1/2} \approx e^2/8\pi\varepsilon_0 Gm_e m_H$, which, using the observed values, yields a number of order 10^{39}. Squaring this gives N, the Eddington number. We note that the number 10^{39} is within a factor of 2 of the ratio of the electrical to gravitational forces between an electron and a proton, and also close to another dimensionless number formed from the fundamental constants, $hc/Gm_p m_e$ – which may be significant or just a curious coincidence.

What is significant about Eddington's work in this area is his discovery of this very large number in nature, namely, 10^{39}. It is the ratio of the electrical to gravitational strengths, and its square is the number of protons in the universe. It is a number that still calls for an explanation. But it is also the ratio of the age of the universe to the time it takes for light to cross an electron, and so the possibility has to be considered that this number is time-dependent since the age of the universe is obviously time-dependent. Arguing against such a time-dependence, Paul Dirac advanced his Large-Number

Hypothesis: any two of the very large dimensionless numbers occurring in nature are connected by a simple mathematical relation, in which the coefficients are of the order unity. If this were a fundamental law, it would mean that the gravitational constant would have to decrease with time, but there is no experimental evidence for this, nor, so far, is there any evidence that any of the fundamental constants is time-dependent. Maybe science is still too young for it to notice. Nevertheless, the possibility of variation is a real factor. But certain quantities could not change because of the way the standards of time and length are defined. Time is defined by the inverse of the frequency of radiation associated with a certain hyperfine transition in caesium, and length is defined in terms of a certain wavelength of light emitted by krypton-86. Whatever change occurs in the fundamental constants that describe these emissions, the quantity obtained by multiplying the krypton wavelength by the caesium frequency, which is a velocity, would not change.

The most interesting consequence of the appreciation of the existence of this large dimensionless number was the creation of a new activity, the 'What If ?' Physics, WIP for short. WIP asked the question, 'What if this number were different from what it is, and why does Dirac's Principle hold so well today?' Answering the second question, Robert Dicke showed that the lifetime of a main-sequence star, T_{ms}, which was determined by its fund of nuclear energy divided by its luminosity (the rate at which energy is emitted as radiation), was dependent on fundamental constants (the gravitational constant, the velocity of light, the proton and electron masses, and Planck's constant). The heavier elements, particularly carbon, on which life is based, are formed only in the late stages of stellar evolution in the more massive stars, and they are spread around the universe by supernovae explosions. Thus, for a universe to support life, its age, T_u, must be at least of the order of T_{ms}. Nor must it be significantly older, since there would then be no stars left to support life as all the stellar nuclear furnaces would have burnt out. Therefore, our universe is such that $T_u \approx T_{ms}$, and it then follows that the Dirac condition automatically holds.

WIP is notable for putting man back at the centre of the universe in the sense that large-number coincidences in physics may thereby be understood to some degree by the fact that we exist. The extreme delicacy of our existence is nowhere brought out more than by Fred Hoyle's analysis of how carbon is formed inside a star. When hydrogen is used up inside a star it is possible, if the star is massive enough, for the ash of hydrogen burning, namely, helium, to become the new fuel, converting to carbon. All subsequent nuclear synthesis, and hence all life, rests on this step. Yet the nuclear physics of our earth-bound laboratories shows that direct conversion is too slow unless helium first converts to berylium, and then berylium plus helium forms carbon. But for this last reaction to take place quickly enough the interaction requires a resonant level in the carbon nucleus close to the energy of the helium-berylium level at 7.3667 MeV. Experiment reveals that carbon has a level at 7.656 MeV, which is close enough for thermal energy to make up the difference. So far, so good, but carbon can be destroyed via collisions with helium to form oxygen. Here again the energies of levels are vitally important. The energy of helium-carbon is 7.1616 MeV and the energy of the oxygen level is 7.1187 MeV, and so, as the oxygen level lies below the helium-carbon level, thermal energy cannot be used to bridge the difference, and consequently carbon is long-lived. On such delicate distinctions is life possible!

Observations, such as the foregoing, that qualitatively the physical structures of the world are consistent with the existence of life reinforce the belief in the unity of the universe. Others have raised this to the dignity of a principle, known as the Weak Anthropic Principle, which states that the observed features of the universe must not contradict the fact of our own existence. Others go yet further and postulate the Strong Anthropic Principle, which states that a universe must be such that life develops and that this is its goal.

This borders on good old-fashioned religion. But we can go another step in anthropic self-image and postulate a Final Anthropic Principle, which, put winningly in objective terms by John Barrow and Frank Tipler, states that intelligent information-processing must come into existence in the universe, and, once it has, it will never die out.[2] However that might be, it is so deeply modest to imply that information-processing is the zenith of achievement of the universe that if I were a universe I would be somewhat miffed (nor much mollified if it were suggested that, somehow, information-processing implied the creation of things like the *Prima Vera*, 'Kubla Khan' and Tristan).

WIP also considers the question, 'What if we lived in a world in which space had dimensionality other than three?' If n is the dimensionality it turns out that the gravitational inverse-square law, $F \sim r^{-2}$, becomes $F \sim r^{1-n}$ and there can be no stable planetary orbits unless $1 < n < 4$. So n can be either 2 or 3. An analysis of wave propagation shows that in two dimensions signals get distorted and only for $n = 3$ do signals travel without distortion. Without the possibility of reliable information from light waves it would be extremely difficult, if not impossible, for us to get any understanding of nature.

It seems that 'we are that we are'; the numbers of nature guarantee it. But we were not always what we are. The same numbers of physics may have always sustained life, and will continue to do so for some time yet, but that is relatively uninteresting to a young, upstart species like us. As far as we know, we are the first to be aware of how delicate our existence is. Exactly who or what we are we just do not know. According to evolutionary biology we are chimpanzees with a slight intellectual edge, mainly because we have somehow acquired an instinct for language which the chimps have not. We are the latest product of a long and incomprehensible evolutionary process that has seen a cooperative of primitive organisms in its native habitat of water grow and change and crawl onto land and grow fur and walk on all fours and learn to stand and lose fur and begin to talk and read *The Times*. We share the achievement, along with protozoa, grass, bees and our cousins the chimps, of possessing an excellently successful set of genes. Our genes, by acting as selfishly as possible, have pulled us through. And here we encounter another number fraught with our destiny, the number of genes in the human genome; a number not precisely known yet – 50,000–100,000 – but it soon will be thanks to the Human Genome Project.

Our genes are our atoms inherited from our parents, who received them from their parents, and they make us look somewhat like our parents and grandparents. Most of them are intent at keeping the body chemistry supplied with the necessary proteins that make us human. Between them and our conscious lives is a vast chemical domain that maintains the well-being of all the colonies of cells our bodies are made of. The genes determine the proteins; physiological chemistry and the environment (diet, trauma, bugs) determine the rest. The degree of determination is analogous to that in a chaotic non-linear dynamic system: the genes may be at the bottom of it all, but there is no predicting the outcome. Just as quarks constitute protons and neutrons, and protons determine the electron cloud, it is the latter that determines the chemical properties. Genes determine proteins and proteins do the business.

Nevertheless, genes acquire a singular importance when they go wrong and give rise to a recognizably inheritable illness, like sickle-cell anaemia. When we hear about genes it is usually in this sort of context, which is somewhat unfair to genes, but good for genetic research, since if a cause of a genetic disease is identified as a wrong sequence of nucleotides on chromosome number 7, say, it somehow holds out a hope of a genetic cure if only more research could be done. Since much of what we know about genes (and a lot of them exist whose function is still mysterious) is in connection with genetically determined conditions, such as the incidence of blue eyes or of the prevalence of Down's syndrome in the children of older mothers, it is natural to suppose that there are one or more genes for intelligence, for language, for athletic ability. But what about genes for determining whether you are a Liberal or a Conservative? whether you like or hate modern music? whether you are religious or not? How much of what you are is nature and how much of what you are is nurture? It is a debate that runs and runs, but now with an increasing emphasis on genes.

We are creatures of our genes as well as of those sonorous numbers of physics. And if the action of our genes is by and large uncertain, we should note that some of those sonorous numbers involve Planck's constant, which is the measure of quantum uncertainty, so both facts give us a measure of how fuzzy we are. Planck originally introduced his now famous constant to explain the frequency dependence of the radiation emitted by a hot body. If radiation energy were emitted in chunks rather than smoothly, as classical physics assumed, and if the chunks were proportional to the frequency, f, of the radiation so that $E = hf$, where h was a constant, then the spectrum could be explained. Without this assumption classical physics predicted an infinite energy output at high frequencies – the so-called ultra-violet catastrophy. Quantum theory was born – energy came in quanta. In the case of the radiation Planck was describing the quanta are known as photons, in the case of sound, phonons. But because all radiation is wave-like, a quantum was spatially fuzzy, its energy spread over a region at least as big as the wavelength. Although particle-like, quanta were very different from the point-particle of classical physics, precisely located. They were fuzzy. Then it was discovered that electrons behaved like waves whose wavelength was related to momentum via Planck's constant. Other particles like protons and neutrons behaved similarly. Matter itself was fuzzy.

So, is the universe such that only fuzzy life develops? Or is it appropriate to apply quantum theory to something as far removed in size as we are, and even our genes are, from the elementary particles whose behaviour gave rise to the theory? It certainly applies to a molecule of DNA. So is there a size limit, or not? The standard view is that quantum effects are universal, but that quantum effects become incoherent in the presence of many interacting particles, so in principle there is no size limit. The effect of incoherence is to make large objects behave according to classical physics. Nevertheless, there are some extraordinarily peculiar properties of the quantum world that suggest to some that they may be connected with that other extraordinarily peculiar property of matter, mind. How plausible is that?

Nine

Substance absolutely infinite is indivisible.

Spinoza, **Ethics**, Prop. XIII

Mathematical physicists are motivated by the vision of Oneness. They are offended by the plurality of particles on the one
hand and by the plurality of forces on the other. Recall that in
the Standard Model of matter there are leptons and quarks and
anti-leptons and anti-quarks. And there are the four interactions: the strong force that binds quarks, with its quantum
the gluon; the weak force that is involved in radioactive decay,
with its own quanta, the intermediate vector bosons; the electromagnetic force, with its quantum the photon; and, weakest
of all, the gravitational force, with its guessed quantum the
graviton. The particles of matter spin with an angular
momentum equal to a half-integral of the fundamental unit
$\hbar$ ($h/2\pi$), where h is Planck's constant, and they can exist
in the same dynamic state with a fellow particle only if
that particle has the opposite spin. As a consequence of this
exclusion principle, they obey the statistics analysed by
Enrico Fermi and Paul Dirac, and they are known collectively
as fermions. The quanta associated with the four forces are
different – they have integral spin and do not obey the exclusion principle; their statistics is that of S. N. Bose and Albert
Einstein, and they are called bosons.

Now, this devilish plethora of fermions and bosons is not
to be borne. Surely, this state of affairs is the result of a Fall.

141 **On** Science

Surely, there was once an Eden where fermions and bosons were merely potentialities within a perfect God particle, the Theon. Surely, the Big Bang was when the Theon exhibited Its glory and created the world. And is it not the reverent and awesome duty of the sons and daughters of Theon to use their Theon-given rationality to trace their evolution back to the Godhead?

But any quest for unity has to bridge the huge gulf that separates the big from the small. Our study of the big has given us gravity, the study of the small, quantum theory. Einstein's General Theory of Relativity has joined together gravity and space-time, and space-time pervades everything, including the world of quantum particles. Yet the two theories – gravitational theory for the macroscopic, quantum theory for the microscopic – are quite separate. This is not acceptable to any red-blooded physicist. And so the search for a theory of quantum gravity has become the most urgent task of modern fundamental physics. Once found, it will be the crowning glory of a study of nature that goes back to the sixteenth century. But it will be physics and not mathematical theology insofar as the new theory predicts new observable phenomena.

This task is not made any easier by the non-intuitive nature of quantum theory. It is a tenet of physics that its laws apply universally, and there is no empirical evidence to suggest otherwise. The spectrum of light emitted by hydrogen is the same here on Earth, in the Sun and in the stars, and that spectrum is explained beautifully by quantum theory. Quantum theory is universally applicable, to events in the furthest star, to events near a black hole, to matter on Earth and to our bodies and genes. Yet the grand paradox of modern physics is that quantum theory cannot account for any fact whatsoever!

The mathematical description of quantum theory goes like this. Every physical system is completely described by a mathematical function conventionally denoted by Ψ and called a wavefunction or a state function. This function is a function of the various dynamic variables that characterize the physical system and it exhibits the probabilities of the results of measurements that may be made on the system. As long as the system remains isolated from the rest of the universe (including any measuring apparatus) it either remains fixed in time or it evolves in a deterministic manner according to the Schrödinger wave equation.

This equation was derived from the classical equation for energy, $E = T + V$, where T is the kinetic energy and V is the potential energy. In terms of momentum, **p** (written bold to denote that it is a vector with x, y and z components), the equation for a single particle is $E = (p^2/2m) + V$, where m is the mass of the particle. Schrödinger converted this into a differential equation for the wavefunction simply by defining **p** and E as differential operators instead of algebraic quantities, specifically, $p_x = -i\hbar\partial/\partial x$, $E = i\partial/\partial t$, where i is the square root of minus one and $\hbar$ (h cross) is Planck's constant divided by 2π. His equation then reads, in one dimension:

$$\left(-\frac{-\hbar^2}{2m}\frac{\partial^2}{\partial x^2} + V\right)\psi = i\hbar\frac{\partial\psi}{\partial t}.$$

This involves a relationship between the time-evolution of the wavefunction and the quantum-mechanical representation of the kinetic and potential energy of the system. If a measurement is made of some dynamical quantity with possible values q_1, q_2, q_3, etc., the wavefunction collapses into the state in which the value of the observable is, say, q_i with probability $|\psi(q_i)|^2$. The way in which this collapse takes place is not described. The appearance of the quantum of action, ℏ, (h/2π), and the square root of −1 in the Schrödinger equation highlights the non-classical nature of the mechanics.

The above formulation describes accurately all non-relativistic quantum-particle phenomena hitherto investigated. Niels Bohr insisted that this gave a *complete* description and that the collapse of the wavefunction was simply the result of attempting to probe the submicroscopic world using macroscopic instruments and employing the concepts of classical physics. Whatever a quantum particle was it simply could not be regarded as a classical particle with definite position and definite momentum. Measurement, by its very nature, disturbed what was being measured. Heisenberg's Uncertainty Principle applied to so-called conjugate variables such as position q and momentum p, so that if Δ denoted uncertainty, then $\Delta q \times \Delta p \geq h$, where h was Planck's constant. It was vain to think of a quantum particle as having a definite position and a definite momentum independent of measurement. The theory was nothing more than a mathematical tool for describing the results of and the relationship between measurements of classical dynamic quantities. To believe that a particle possessed these quantities before an observation was made was unwarranted. This pragmatic and positivistic approach became known as the Copenhagen Interpretation.

Einstein never accepted this. For him the whole point of physics was to describe the mechanical world in terms of the deterministic motion of particles under the local influence of forces. It made no sense to think of a particle that did not have a definite position in space and a definite momentum at a particular time. Quantum theory was therefore incomplete in that it could not describe the dynamics of particles in that way. Insofar as it was successful in describing submicroscopic phenomena it had to be regarded as describing populations of particles and not single particles. In other words it was a statistical theory, the statistics, as usual, cloaking ignorance of the detailed mechanisms operating submicroscopically.

Quantum theory saw an electron as a wave-packet spread in space with a range of wavelengths rather than as a classical particle. The double-slit experiment, familiar in laboratory demonstrations of the wave nature of light, became a paradigmatic empirical demonstration of the wave-like nature of the electron. A beam of electrons prepared so that each electron had the same velocity, and therefore the same momentum and therefore the same wavelength, is incident on a screen in which there are two slits. Beyond the screen is another coated with a phosphor that emits light when an electron hits it (just as happens in a television set), and this shows an interference pattern. A statistical theory might suggest that the pattern arose not from any wave-like nature of individual electrons but from electron–electron interactions of some sort. Experiment shows that the interference pattern builds up even when the intensity of the electron beam is reduced so that on average only one electron at a time is incident on the two slits. The same experiment can be done with light when only one photon at a time is present. Electrons and photons share the same seemingly intrinsic wave-like nature, and yet behave as particles when detected. In our macroscopic experience nothing can behave in such a mutually contradictory way.

A famous thought experiment was suggested by Einstein, Podolski and Rosen (hereafter EPR) in 1935 in order to support the claim that a complete theory must contain every element of physical reality and therefore quantum theory was incomplete.[1] The meaning attached to the idea of an element of physical reality was that it corresponded with a physical quantity that could be predicted with certainty from the results of other experiments without further disturbing the system. They considered a system of two particles that interact at $x = 0$ and fly apart with equal and opposite momentum, ceasing immediately to interact. After some time the momentum of particle 1 is measured to be p_1; then the momentum of particle 2 can be deduced with certainty to be $- p_1$. Another experiment is performed, but this time the position of particle 1 is measured to be x_1; the position of particle 2 can be deduced with certainty to be $- x_1$. Thus both momentum and position have been shown to be elements of reality associated with particle 2 which have been established without disturbing particle 2. But quantum theory denies that such elements exist and points out that two distinct experiments were required. EPR, foreseeing this objection, pointed out that the implication here was the claim that simultaneous elements of physical reality are only so if they can be simultaneously measured or predicted. If this is the case then the EPR argument fails, since it is certainly the case that either momentum or position can be measured, but not both simultaneously. EPR then dismiss this definition of reality as unreasonable because the reality of the position and momentum of particle 2 would then depend on what measurement was carried out on particle 1, however far apart the two particles were. Therefore quantum theory is incomplete.

The argument of EPR and their conclusion rest on the assumption that once separated the two particles cease to interact. This assumption turns out to be invalid – undisturbed quantum systems are intrinsically holistic and nonlocal. Thus a measurement of the momentum of particle 1 instantaneously determines the momentum of particle 2, but no definite position can be assigned to either particle. No meaning can be attached to momentum and position until a measurement of one or the other takes place. This property of non-locality has been amply confirmed in the laboratory even over macroscopic distances of the order of a metre. The wave–particle duality of matter can perhaps be assimilated without overmuch conceptual trauma by contemplating the properties of a classical wave-packet, but the holistic, non-local nature of quantum systems has no classical placebo to relieve the shock.

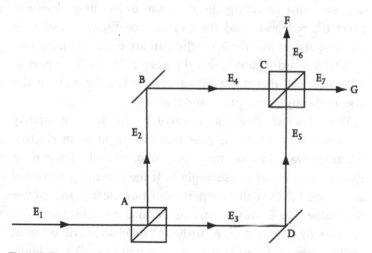

Figure 4: The interferometer

A straightforward modern demonstration of non-locality can be made using the interferometer shown diagrammatically in Figure 4 (that is, staightforward conceptually – nothing is straightforward in practice, as any experimental physicist knows). A and C are beam-splitters and B and D are 45° reflectors. A collimated beam of monochromatic light enters from the left, is split into two parts at A and after reflection at B or D split again at C. The output is detected at F and G. In general, what is observed is interference patterns that depend upon the difference between the lengths of the arms ABC and ADC, the size of the reflection and transmission coefficients, the phase changes in reflection and transmission, and any loss processes like absorption and scattering.

In order to simplify things let us assume that we have an ideal instrument for our purposes. This means that the interferometer is such that there is no difference in path length between the arms, that the beam-splitters are symmetrical in the sense that reversing the direction of the light does not affect the reflection and transmission coefficients, and they are designed so that these coefficients are equal in magnitude (a 50/50 beam-splitter), that the mirrors B and D are perfect reflectors, and that not a bit of light is lost throughout the system through absorption and scattering.

The classical description, which, in fact, accurately describes what happens, goes like this. Light is an electromagnetic wave whose amplitude is described in terms of electric field E and phase angle ø. If the incoming wave has amplitude E_1, then the properties of our ideal beam-splitter determine that E_2 and E_3 have identical amplitudes but differ in phase by exactly $\pi/2$. A further phase change will occur at the two mirrors B and D, but since the two mirrors are identical the phase difference between E_4 and E_5 remains $\pi/2$. At C a further difference of $\pi/2$ is introduced between the reflected and transmitted beams. In the case of E_6 this extra phase difference adds and so the reflected part of E_4 is out of phase with the transmitted part of E_5 by π. This is the condition for destructive interference and so no signal is registered by F. In the case of E_7 the extra phase difference subtracts so that the transmitted part of E_4 and the reflected part of E_5 are in phase. This is the condition for constructive interference, and G registers a signal. Classical wave theory explains what happens in an entirely satisfactory way.

Now consider the experiment from the point of view of quantum theory. We know that electromagnetic energy is transmitted in packets called photons. The classical result of the experiment can be envisaged as arising from interference between photons one of which follows the path ABC, the other, ADC. This can be tested by reducing the chance of more than one photon at a time being in the interferometer by reducing the intensity of the incident light. Unfortunately, the chaotic production of photons from an ordinary source of monochromatic light makes the required condition hard to obtain in a convincing way, so special techniques, which we need not describe here, have to be used to ensure that indeed only one photon at a time goes through the apparatus. Building up a significant result after many single photons have passed through leads to agreement with the classical result. Interference is a single-photon effect.

This phenomenon is incomprehensible if the photon is regarded as a classical particle. It is comprehensible only if the photon somehow travels both paths at once. Yet if a detector is inserted in one path it will now and again intercept a single photon, but the presence of the detector destroys the interference, which means that the average output of F equals the average output of G. Clearly the photon interacts holistically with the measuring equipment. Provided the paths remain isolated from the rest of the environment the interferometer can be as big as one would like. The astonishing thing is not only the fact of quantum non-locality but also the fact that it is not restricted to microscopic dimensions.

Equally shocking was the idea that the act of observation determines the elements of reality. This seemed to bring the whole idea of an objective reality into question. It certainly has modified our picture of the world, but not to the extent that any radical subjectivity has been admitted, though early analyses of the collapse of the wavefunction tended unhappily to overemphasize the rôle of the mind.

An early theory due to von Neuman and to Wigner connected one mystery with another and proposed that the collapse was triggered by the mind of the observer. This was a return to magical ideas with a vengeance. Imagine a cat in an opaque box in which there are a weak radioactive source and a deadly poisonous gas that is released once a radioactive decay (which occurs at random) is detected. The correct quantum-mechanical description of the cat under these circumstances is a wavefunction that is a linear superposition of two wavefunctions, one for the cat alive, the other for the cat dead. Now an observer lifts the lid of the box and sees whether the cat is alive or dead. Since the cat cannot be both alive and dead, the wavefunction immediately collapses to one or the other component. The thought of a cat in a quantum-mechanical state that was a superposition of being dead and being alive troubled Schrödinger quite naturally, and it is even more troubling to imagine collapsing the wavefunction into either dead or alive simply by looking. Unbelievable!

The question arose when exactly does the wavefunction collapse in a measurement? When the pointer on the measuring apparatus registers a definite reading? or when the experimenter notes the reading? or when the experimenter's supervisor reads the notebook? or when the paper is published and read by the scientific community? I dare say the last answer would appeal quite reasonably to the interested, but surely by now incredibly naive, sociologist.

The part played by measurement, and by implication an observer, posed and continues to pose a serious problem. We are intimately familiar from childhood, at least qualitatively, with the behaviour of the world of classical physics. The classical world is the real world. But there is also the submicroscopic world described accurately by quantum theory of which all the matter of the classical world consists. Somehow the classical world must be seen to emerge naturally from the quantum world. But in terms of the present theory it does so only through measurements in which the holistic nature of the quantum world disappears with the collapse of the wavefunction and a definite event occurs.

But definite events occur without any measurement process taking place. This is especially a problem for astrophysics and cosmology, there being no observers all over the universe and certainly none at the Big Bang from which galaxies formed. Clearly, a quantum cosmology cannot contain the paraphernalia of measurement-induced collapse of the wavefunction. One extreme approach to this problem was taken by Hugh Everett III supported by John Wheeler. It could have been dubbed the Princeton Interpretation to replace the Copenhagen Interpretation but, instead, it became known as the Many Worlds Interpretation. The nub is the jettisoning of the collapse of the wavefunction. The wavefunction is given primary ontological significance with the Schrödinger equation describing its evolution for all time, and all the potential events which it contains exist for all time. There is just one universe. Somehow we, as observing minds, become aware of one, and only one, evolving branch of this Universal Wavefunction no bits of which, including the bit we are aware of, ever collapse. Other sets of minds could be aware of other evolving branches, but there cannot be any communication with them. This theory is not so much physics as metaphysics. The universe out there is an objective universe that exists independently of us but we can be aware of only one aspect of its evolution. Sometimes this is called the Many Minds Interpretation.

There are few physicists who feel comfortable with Everett's theory. They are bothered by its psycho-physical flavour. Nevertheless, the idea of abandoning the concept of a collapsing wavefunction remains attractive in the cosmological context. DeWitt's version of the Many Worlds theory is entirely objective. He assumes that interactions of a measurement-like nature continually occur and at some point in the interaction the universe actually splits, each branch consisting of one of the possible outcomes of the interaction. We are part of one of those branches. What determines the point at which the splitting occurs remains a problem.

Gell-Mann and Hartle have recently elaborated this approach by evoking the idea of histories determined initially at the time of the Big Bang.[2] If a fine-grained view of the universe is taken, then these histories cohere and exhibit all the interference effects and potential events as described by the usual quantum theory. At a coarse-grained level they assume that the histories can decohere and manifest this decoherence in terms of actual events that can be described quasi-classically. Basically, they assume that conventional quantum theory breaks down at the course-grained level and holds only at the fine-grained level. What size the graining must be for this to happen is still a problem, but it may not be a universal size – it may depend on the physical situation.

It is evident that all these theories that kill the collapse of the wavefunction have to resurrect it in some way or another – in terms of mental awareness, or actual branching, or coarse-grained decoherence. Another approach is to take the collapse of the wavefunction as a real physical event and to attempt to model it. One of the earliest ideas was that of David Bohm, who assumed that any interaction with a measuring device, inevitably large-sized and irredeemably classical, introduces random phase-factors into the wavefunction as a result of the interaction modifying the Schrödinger equation.[3] Averaged over many measurements these random phase-factors would destroy the wavelike interference characteristic of quantum systems, leaving only the wave intensities that determine the classical probabilities.

Another, more radical, approach to explaining the collapse was that of Ghirardi, Rimini and Weber. They assumed that Schrödinger's equation was incomplete and they introduced a term that describes a spontaneous localization of a wave into a region of dimension $0.1\mu m$ at a rate of once every $10^{16}s$ (roughly once every 10^9 years). This process would be virtually undetectable in a microscopic system but in a macroscopic system, with 10^{23} particles per cubic centimetre it would be effective. In principle this idea is testable, but in practice, at least with today's capabilities, it is not.

A more recent idea is to associate the collapse of the wavefunction with gravitational effects. Roger Penrose points out that alternative possibilities embodied in a branching wavefunction implies different distributions of mass and therefore, via the connection in general relativity, different space-time geometries.[4] The bigger the mass involved, the bigger the effect and the sooner a definite choice is made – surely the world cannot cope with a plurality of space-times. This is to associate a quantum effect with an effect in general relativity.

Unfortunately, the problem of incorporating quantum theory into the theory of general relativity is still intractable. We have already mentioned the problem that a vacuum bubbling with virtual particles poses. Since we do not have a quantum theory that incorporates general relativity, Penrose's remark cannot be fleshed out.

It has to be said that the prevailing theoretical climate of opinion is that the collapse of the wavefunction is no more than a convenient summary of what really goes on at the interface of quantum and classical systems. No such collapse truly occurs; the reality is much more subtle, demanding an extension of quantum theory itself.

All the interpretations of quantum theory, including the Copenhagen one, describe quantum phenomena in terms of more-than-usually abstract mathematics that provides rules for handling experimental data. What lies behind these phenomena is regarded as inconceivable. The instrumental approach, exemplified by Stephen Hawking,[5] is to say that we have a perfectly good mathematical theory, so what more do you want? This attitude is totally at odds with that of classical physics insofar as it sees things as existing in themselves and moving along definite paths in response to well-defined, locally acting forces. But, actually, the instrumentalist view is no more or less than that adopted by Newton long ago in his mathematical description of gravity – the effects of action at a distance are described but not explained. Nevertheless, the Ptolemaic theory of planetary motion, based on the Earth being at the centre, was wonderfully accurate, but nobody believes it to be a true representation of reality, not even professed instrumentalists. As I have pointed out, the quantum attitude was anathema to Einstein, who believed that it was not what physics was about. This was a view that directly influenced David Bohm, who developed a 'hidden-variable' model that is still viable today, in spite of the many criticisms it attracted when it was first published in the 1950s.

The basic ontology of Bohm's approach is not the wavefunction, as it is for all other theories, but the particle pursuing a classical trajectory, a trajectory, however, that was determined by a non-classical field characterized by a quantum-mechanical potential, Q.[6] Bohm relates this potential to the wavefunction and its spatial dependence and then shows that this new potential just adds to the ordinary classical potentials of gravity or electromagnetism to determine the motion of the quantum particle through the Hamilton–Jacobi equation of classical physics. Indeed, the condition for entering the classical regime is that Q be negligible. When Q is not negligible the motion of the particle is determined by the quantum Hamilton–Jacobi equation and the wavefunction evolves according to Schrödinger's equation.

The essence of Bohm's approach is to regard a quantum particle as following a classical trajectory in the presence of, in general, two fields, one the usual classical field if present, the other, always present, a new quantum-mechanical field. All quantum phenomena, including interference and non-locality, can be accounted for. The approach has a strong affinity with the idea of Louis de Broglie that the wave introduced by quantum theory was a pilot wave that guided the particle. In Bohm's theory it is the quantum potential that determines the motion, but the quantum potential is itself defined by the wave, which is further defined by the whole of the experimental set-up. Thus in the two-slit experiment Q has a spatial pattern that forces an electron, which now definitely goes through one slit or the other, into exactly those directions that lead an ensemble of electrons to exhibit an interference pattern.

Bohm's theory, sometimes called the Causal Interpretation, can also explain the apparent collapse of the wavefunction in a measurement. The meaning of a measurement is not the classical one of observing a property of the object alone but, as in the standard interpretation, depends on the potentialities of the combined system of object and measuring device. During the interaction the various components of the wavefunction will overlap and interfere and the quantum potential will be very complicated, offering a number of channels for the particle to enter. Depending on the starting conditions of the particle one channel will be selected and this will change Q. If only a few particles are involved, inactive channels can still overlap and the effect can be reversed. But if the measuring apparatus is a macroscopic system containing a huge number of particles, as is the usual case, the complexity and multiplicity of the channels will be so large that subsequent overlap is extremely unlikely, and the choice of channel becomes irrevocable. (This explanation recalls Bohm's idea of random phase-factors.) No collapse of the wavefunction is involved: real events occur because the trajectories of quantum particles are determined by initial conditions and by the disposition of macroscopic objects that determine the quantum potential.

The picture of the world that emerges from the interpretation of quantum theory by Bohm and others is of a holistic quantum substratum of immense complexity that underlies a familiar world of apparently distinct individual objects. Every bit of this quantum substratum contains information about the whole, in the same way that any bit of a hologram contains information about the original object from which the hologram was formed. In Bohr's terminology the quantum world constitutes an implicate order that becomes manifest, explicit, in classical-sized objects. The connectivity of things in the world is an old magical idea. Good old ideas, it seems, never die!

Nevertheless, we are presented here with something very odd in theoretical physics. There is a tremendous conceptual rift between the utilitarian interpretation of Bohr and all other interpretations. Insofar as the Copenhagen Interpretation limits itself to an Occam's razor view of what can be known, it is an epistemological theory – the wavefunction and the dynamic variables have no objective reality. Other interpretations are ontological in that they grant an objective reality to the wavefunction, and out of this all the problems connected with wavefunction collapse arise. The driving belief here is that quantum theory should be a theory of mathematical physics that is applicable everywhere in the universe and out of which the familiar classical world should emerge naturally. In a highly mathematico-logical approach, exemplified by d'Espagnat[7] and Omnès,[8] the quantum theory is based on a set of axioms, from which the physical behaviour of the universe is to be deduced. Such a scheme must provide a clear account of the collapse of the wavefunction since that is what is empirically observed. But the collapse has to be only an apparent collapse, and here we have to include *interpretation* in the theory – Many Worlds, Many Minds, Decoherence, Causality, whatever. The opportunity for mathematical sophistry on this matter seems endless, and, one may add, pointless scientifically unless real experiments are suggested that provide tests. Until a programme of such experiments is under way, it is a matter of faith and inclination which *interpretation* to adopt.

Experiments do go on, but whether their results can ever influence interpretation may be doubted. The huge problem that quantum theory has is that *it cannot explain the existence of facts*. As long as the theory consists only of statements concerning probabilities it cannot contain an explanation of the physical features of the world as described by classical physics, and of the world as we know it through our impressions and experience. Things do happen – this thing rather than that thing – and we need to understand why. We have a magnificent theory that describes the coherent dynamical behaviour of microscopic systems – what we lack is a comprehensive theory of decoherence that explains the emergence of macroscopic events. The causal theory of Bohm has the virtue of relegating statistics to the same domain that includes classical statistics, that is, a domain that exists because of our ignorance of the detailed behaviour of deterministic trajectories. The definite result of a measurement made on a microscopic system is, in principle though not in practice, traceable back to the starting conditions. If some are attracted to the Causal Interpretation, many others prefer to imagine Many Worlds, with decoherence somehow leading to definite events in at least one world. It seems to depend upon temperament.

But nothing sharpens the scientific mind more than the possibility of technological exploitation. Whatever the problems of interpretation, quantum theory is in everyday, down-to-earth use. The transistor, the laser, the nuclear-magnetic-resonance brain scanner cannot do without it. As long as the transistor is more than a few hundred atoms long, an electron going from source to drain, to all intents and purposes, behaves like a classical particle, albeit with a somewhat weird relation between its momentum and energy. Decoherence is rapid, occurring in times as short as a hundred femtoseconds ($1fs = 10^{-15}s$), and this means that the electron is almost always in a well-defined state. In a laser the inevitably present forces of decoherence are swamped deliberately to produce light in a pure coherent state. In nuclear magnetic resonance (NMR) the nucleus is so well insulated from the rest of the universe that magnetic states within the nucleus can live coherently for milliseconds or even seconds.

The study of the mechanism of decoherence is of prime importance for understanding the behaviour of these various devices. But these devices are not simple systems. They are macroscopic, involving vast numbers of atoms. In order to study how coherence leaks away into the rest of the universe much simpler systems need to be studied, and this fundamental investigation is under way in many laboratories. Whether the results of such investigations will bear on the problem of how quantum theory is to be interpreted remains to be seen.

Decoherence is not a problem for the transistor, the fundamental unit of the modern digital computer. A computer uses binary digits – bits, in the jargon. A bit is represented by an electronic component that is either on or off, the on state, for example, representing 1, the off state, 0. Without decoherence it would be quantum mechanically possible for the component to be in a mixed state, equally on and equally off. This would never do, and it does not happen with conventional electronic components. The worst that can happen is for an error to occur that makes it a 0 when it should be a 1, or vice versa, but there are extremely sophisticated error-correcting techniques that have been developed which deal with this. The possibility 0 or 1 is bad enough, but 0 and 1 would be a nightmare.

Or would it? An enthusiastic group who have become

more prominent in recent years with messianic members is the group working in the field of quantum information, which includes quantum computation, quantum cryptography and quantum teleportation.[9] It is based on the existence of the qubit, the quantum mechanical equivalent of the bit. Many have responded to the call of the qubit with fiery enthusiasm. A qubit is any two-level system that can be placed in a mixed state for long enough for it to become entangled with other qubits and for various manipulations to be carried out before decoherence occurs. The basic attraction is that whereas an ordinary bit has two states, 0 or 1, the qubit has four states, a, b, a + b or a − b, the last two being the mixed state whose sign reflects the effect of interference, + if the interference is constructive and − if it is destructive. It opens up the possibility of parallel computation, which, in principle, allows computational problems to be solved that are too time-consuming by ordinary means.

One use of a quantum computer would be to factorize N-digit numbers, where N is an integer, into its primes in a time that increases as N^3, which is much better than the case for a conventional computer where the time increases exponentially with N. This facility would have its application in the field of cryptography.

This is a good moment to introduce the protagonists Alice, Bob and Eve, without whom no account of quantum information would be complete. Alice and Bob are upstanding characters who appear to be in perpetual communication, though always at a distance. Eve is a rather shady person who is keen to find out what they are saying to one another, possibly because she is jealous, but of which one it would be politically incorrect even to hazard a guess.

One of the standard ways Alice and Bob could use to

On Science

communicate secretly is for them to share a secret key to decode what they hope looks like gobbledegook to Eve. The trouble with this system is that if it is used all the time Eve will ultimately work out what the key is, so, ideally, it can only be used once. This implies that the secret key for the next message has to be mutually communicated in some way, but this could perhaps be intercepted by Eve. A better way is to use a public key in the form of a huge number whose prime factors are known to, say, Bob (which is how he worked out the huge number in the first place). Alice uses the huge number to encrypt her message and Bob uses his factors to decrypt it. Eve has to struggle with the probable fact (though no proof for this) that the time to factorize an integer increases exponentially with N. This is where the factorizing algorithm using a quantum computer comes in, reducing the time dependence from e^N to N^3. So Eve could use a quantum computer, though it would have to be a big one.

This would force Alice and Bob to turn to quantum cryptography. The details are straightforward but technical. Suffice it to say that the use of entangled qubits allows Alice and Bob to share a secret key chosen quite at random, though physically separated, without the need of a courier or a meeting. And, moreover, they will know whether the key has been intercepted because any measurement made on a quantum system destroys the coherence.

Sometimes Bob has a passion to know something about Alice's qubits. Suppose Alice has a qubit whose state is unknown to Alice but she knows that Bob would like to find out for himself. She knows that if she just went ahead and measured the qubit's state in order to tell Bob, that would destroy it and Bob would not have the fun of finding out himself. Happily, they are each in possession of one qubit of an entangled pair which has magically avoided decoherence. Alice now entangles the unknown qubit with her member of the pair and she makes a measurement which collapses the total entangled state. She then sends the result of her measurement to Bob, who can then deduce what operation he must apply to his qubit to convert it into one with exactly the property of the unknown. Quantum teleportation means really the transmission of quantum information using the non-local nature of entanglement.

There are many other schemes where the odd properties of the quantum world can be exploited in principle. There are even ingenious schemes whereby errors can be corrected without a measurement having to be made on the qubits carrying the information. However, theory, as in many cases, is way ahead of practice. Experimental implementations do exist, but they typically involve no more than two qubits. The problem of avoiding or correcting for decoherence in the message and in the error-correcting processes themselves is extremely intense. Serious quantum computation is not envisaged for some time to come, and many think that it will never occur. But nobody who is enthusiastic has ever listened to the wet blankets, so research in this area will continue unabated.

The call of the qubit is all around us, but if few hear, it may be because of its highly technical sound. Those qubits involving the spin of the electron or the polarization of the photon involve the effects of special relativity. The qubit is, nevertheless, a technological creature of quantum theory that is being put to use, if only, as yet, in a limited way. In this field decoherence is a technical problem rather than a scientific one; something to be avoided or weakened wherever possible. That is the major problem. Something may turn up to make the whole thing viable – maybe some new physical principle. But it is as desperate as that. In the end the principal effect may be that Alice, Bob and Eve achieve nothing but a quasi-literary status.

Extending quantum theory into the realm of special relativity in the form of quantum-field theory, which fills all space, even empty space, with the bubbling activity of spontaneous creations and annihilations of particles and anti-particles, leads to more problems (like infinite energies). Extending it further to include gravitation is as yet impossible. And without all of that, what price premature Universal Cosmological Theories of Everything, the Big Bang, The Beginning, and the Mind? Science here has a long way to go, but, in the process, it has to be wary of transforming into pure mathematical theology on the one hand, and on the other, straying into confusion, which occurs, according to Bacon, 'wherever argument or inference passes from one world of experience into another'. The discovery of quantum phenomena has made most physicists cannier about applying their knowledge beyond their own subject. Nature is much more subtle than the creators of classical physics imagined, and there may well be more surprises in store. But for some, it is irresistible to yoke one mystery with another, and one field into which science strays at its peril is the mind.

Science and the Mind
Ten

What is matter? – Never mind.
What is mind? – No matter.

Punch 29 (19) (1855)

The problem of the relationship between mind and body has occupied philosophers from the beginning and is likely to continue to do so into the forseeable future. A rough summary of the ideas about the mind that have been advanced is depicted in Figure 5. Recently, the problem has re-emerged in science itself, largely triggered by advances in computer technology, which has set off lively, not to say passionate, discussions in the scientific literature. But whether science has anything meaningful to say on the topic, particularly where consciousness is involved, is certainly not clear. Science can, however, offer a definition that puts mind to some degree within its scope. Thus, the naturalistic view of mentality is that it is a feature of the enormously complex interactions that people have with the natural and social environment, a feature that has evolved over time. Without being an out-and-out dualist it is difficult to see anything wrong with that.

Nevertheless, science aims to develop a view of nature that is coherent and self-consistent, and most scientists would be reluctant to limit its scope merely to the universe of inanimate matter, nor do they in fact. Animate and inanimate objects are equally proper material for scientific investigation, and what is found to be true of one category must be at least

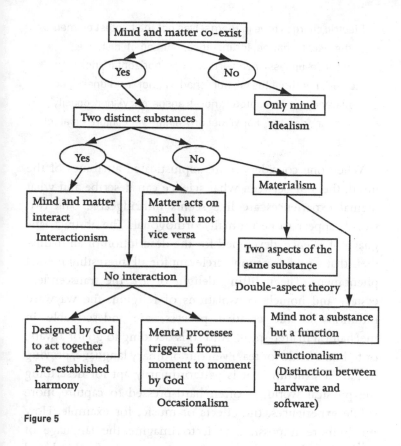

Figure 5

compatible with what is true of the other. But here the science of matter runs into a conceptual brick wall. Living, conscious objects inhabit a world that is inaccessible, a world in which the very language of physical science is inapplicable. Even putting aside glaringly obvious disconnections like value and morals, none of the down-to-earth sensations – which even scientists experience – such as the act of vision is remotely adequately describable. As Sherrington has put it:

Electric charges have in themselves not the faintest elements of the visual – having, for instance, nothing of 'distance', 'right-side-upness', nor 'vertical', nor 'horizontal', nor 'colour', nor 'brightness', nor 'shadow', nor 'roundness', nor 'squareness', nor 'contour', nor 'transparency', nor 'opacity', nor 'near', nor 'far', nor visual anything – yet conjure up all of these.[1]

When one considers more sophisticated attributes of the mind, the gulf between what science can describe and what mental experiences are like seems unbridgeable. The two worlds appear to be mutually orthogonal. The abstract language of science, developed for the description of an if-this-then-that world, is simply irrelevant for all interesting mental phenomena. The uplifting delight of art, the transcendent ecstasy and homely consolations of religion, the wayward intuitions of the creative process, are indescribable in mechanical matter-speak. And so is deciding to go for a walk, or to boil a kettle for tea. Even our ordinary language, proving to have a structure that is extraordinarily apt for describing the physical world, is often hard-pressed to capture more subtle experiences, the effects of music, for example. How much more impossible is it to imagine the language of mathematical physics applied to the world of ethical and aesthetic value!

But if the life of the mind is beyond the powers of science, does this mean that we have to accept a mind–matter duality? I think that in one sense the answer to that is obviously yes, in another sense, no. Clearly, mind appears to us to be so totally different from matter that it will continue to regard itself and its concerns as in a quite distinct category. Behaviourist ideas that regarded the concept of mind as a category error seem now to be very dated. The problem of how consciousness can arise from matter is currently a hot topic. Mind is inseparable from matter since it can directly manipulate the body and be affected by the body chemistry. It is therefore both spiritual and material, so if the spiritual activities of the mind – aesthetics, ethics, religion (and we might add sensation) – are beyond the scientific pale, the brain is not. This distinction between these two attributes of mind – the software and hardware, in computer-speak – is often used for comprehending and assessing the efforts made in recent years to link the mind with the computer and quantum mechanics.

Roger Penrose usefully distinguishes four modern views about a scientific theory of mind:

A. All thinking is computation; in particular, feelings of conscious awareness are evoked merely by the carrying out of appropriate computations,

B. Awareness is a feature of the brain's physical action; and whereas any physical action can be simulated computationally, computational simulation cannot by itself evoke awareness.

C. Appropriate physical action of the brain evokes awareness, but this physical action cannot even be properly simulated computationally.

D. Awareness cannot be explained by physical, computational or any other scientific terms.[2]

View **A** is identified by Penrose with the hard-line approach of many in the field of artificial intelligence (AI), namely, strong AI. If a machine passes the famous Turing test, that is, if it behaves completely as if it had intelligence, it actually has consciousness. It is interesting to consider the converse of this. The claim would be that if an animal passes the test for consciousness, it actually has intelligence. Intelligence is presumably the ability to solve problems. Not all animals are gifted with this ability, but few would deny that they are conscious. Certainly, it cannot be claimed that the ability to problem-solve and consciousness are indissolubly linked, that one implies the other. To claim that the implication is unidirectional seems rather weak.

Even less persuasive is the idea that the machine understands what it is doing, understanding presumably being one of the attributes of the machine's being conscious. There is a yawning gap, in other words, between semantics and the syntax of the computer program. In his Chinese room argument, John Searle, knowing no Chinese, has an image of himself isolated in a room, but equipped with a set of rules (in English) for manipulating Chinese symbols that tell a story fed to him through a slot. Without knowing a word of Chinese, but following all the rules of manipulation, he could correctly answer questions put to him about the story even though he would not have a clue what the story was about. There are many human contexts where the ability to follow instructions is by no means matched by understanding. The fact that machines are excellent at carrying out instructions simply does not imply the existence of understanding, or indeed of anything relating to consciousness.

The gap between semantics and syntax, so graphically illustrated by the Chinese room, is enough to destroy the claim of strong AI. But worse for strong AI is to come. Searle has pointed out that, in addition to the gap between semantics and syntax, there is an even more potent gap between syntax and physical processes such as the firing of neurons.[3] In a computer a transistor might be on or off depending on what electrical signals it has received. The computational syntax in the mind of the programmer is that off means 0 and on means 1. The syntax is certainly not in the transistor and its properties, nor can such syntax reside in brain cells. Bridging the gap between firing neurons and conscious understanding needs a much more sophisticated approach than is provided by computer-speak.

Awareness as a feature of the activity of the brain is not denied by any of the four views. To do so would be either to espouse a full-blooded idealism in which the world, if it exists outside the mind, is one of appearance only, or to believe in the soul, the deus ex machina, quite separate from the body. It would be vain to add to the literature on these dispositions – Kant on the one hand and the religious books on the other are sufficiently exhaustive. Also Byron:

> When Bishop Berkeley said 'there is no matter',
> And proved it – 'twas no matter what he said.

I incline to the metaphysical materialistic view that nature is sufficiently subtle for matter to give rise to mind somehow, so for present purposes I intend to stick with the Penrose four.

View **B** is the soft-AI position which does not see awareness arising out of mere computation, even though it assumes that all physical action can be simulated by a computer. Soft AI is unobjectionable from a methodological point of view in that it offers the possibility of designing computer models of certain brain activities that may have relevance to the science of the brain. This would assume that the brain functions formally in the same way as a computer does, which may, or may not, be correct. If indeed the brain functions in a formal way then the approach is justified, since a computer can simulate any formal system, even if it does not 'understand' anything.

View **C** denies that all physical action can be computerized, and in particular the physical action that leads to awareness. View **C** is the view of Penrose as developed in his two books, *The Emperor's New Mind* and *Shadows of the Mind*,[4] and this has provoked energetic responses from the strong-AI community, from Nobel prizewinners and from philosophers, some of which are sadly intemperate. A few quotations will give the flavour:

According to Horgan in his book *The End of Science*,

> Minsky [AI] called Roger Penrose a 'coward who could not accept his own physicality'.[5]

Nobel prizewinner Phil Anderson reviewing *Shadows of the Mind*:

> I was put off reading *The Emperor's New Mind* by the many critical reviews it received, so I came to the sequel fresh, albeit prejudiced. Let me say without hesitation that my prejudices have been amply confirmed.[6]

Nobel prizewinner Murray Gell-Mann:

> Roger Penrose has written two foolish books based on the long-discredited fallacy that Gödel's theorem has something to do with consciousness requiring – *something else*.[7]

Philosopher Hilary Putnam reviewing *Shadows of the Mind*:

> And yet this reviewer regards its appearance as a sad episode in our current intellectual life.[8]

Minsky would presumably brand anyone who held a view different from **A** in the same way. Anderson thinks that Penrose should by now have thoroughly accepted Gell-Mann and Hartle's version of the collapse of the wavefunction and should not have brought in general relativity to explain it. Gell-Mann thinks that Gödel's theorem is inapplicable, a view shared by Putnam whose response as a philosopher probably stems from exasperation – surely it is evident that mind cannot be described by physics. Whatever one may think of Penrose's ideas, his books have triggered off a tremendous debate notable more for its aggressiveness than its usefulness.[9]

The nub of Penrose's case against hard and soft AI is that not everything is computable. Rolf Landauer of IBM would retort that what can't be computed is meaningless. In this extraordinarily reductionist view the world is understandable only in terms of binary digits. But, says Penrose, there are real mathematical structures in the world – Platonic forms, no less – that cannot be simulated computationally. A computer works from a set of rules – an algorithm. Penrose argues that mathematical insight, understanding, meaning and judging whether anything is true or not are functions of the brain that are non-algorithmic. Learning programs for computers are still algorithmic and so could never teach a machine to understand. No set of rules, however sophisticated, could be complete enough to encompass all the insights into nature that are possible to the human mind. In support of this claim he cites Gödel.

We have already had reason to mention his famous theorem in our discussion of mathematics. It seems that the suggestiveness of Gödel's proof for models of the mind, as well as for models of nature generally, is infinite, but unfortunately the applicability is not. Lucas was one of the first to claim that Gödel's theorem proved that minds could not be explained as machines. The theorem states that in any consistent system which is strong enough to produce simple arithmetic there are formulae which cannot be proved in the system, but we can see that these formulae are true. Within the system we can always discover a formula **G** that says '**G** is unprovable.' If the formula is nevertheless provable within the system we have a contradiction: the system is inconsistent. On the other hand, if **G** is, in fact, unprovable within the system, the system is incomplete. Thus all consistent formal systems like arithmetic are necessarily incomplete: there will always be formal truths underivable from the inevitably finite set of axioms that define the system that we can perceive to be true. Because computers are algorithmic they can never incorporate all the truths that a mind can perceive.

One counter argument to this is that almost the whole of mathematics can be represented in Zermelo–Fraenkel set theory, which, as it happens, cannot obviously be seen to be consistent. Thus Gödel's theorem does not obviously apply, and hence it does not obviously apply to the description of the mind insofar as the mind can be modelled by set theory. This may be so, but computers as we know them work on arithmetic, which is known to be consistent, so Gödel's theorem would apply. That, of course, leaves open the question as to whether mathematicians' minds are consistent and whether there is a limit to their insight. If inconsistent, Gödel's theorem does not apply; if consistent but limited a mind-modelling algorithm with a finite number of axioms could be envisaged. Under these conditions the claims of Lucas and Penrose collapse.

View **D** asserts the incompetence of the power of science regarding mind. Penrose rejects this on the grounds that 'it has been only through the use of the methods of science and mathematics that any real progress in understanding the behaviour of the world has been achieved'. So much for art, history and philosophy! But, of course, he means the physical world. But then, what physical action evokes awareness? We are back to this fundamental question. In common with many mind-modellers he succumbs to the blandishments of quantum theory. The fascinating property of quantum systems to exhibit coherence and non-locality sometimes over macroscopic distances has certain analogies to the function of the brain. Thoughts are not here or there: they are non-local. They seem to emerge from a substratum of hazy proto-thoughts in a way that suggests the collapse of some cerebral wavefunction. Modern quantum ideas point to a non-locality that stretches throughout the universe. The potential for magic and mysticism at this point is enormous. Penrose has nothing to do with the wilder fringes of quantum-collapsing minds, but points tentatively towards elements in the brain where quantum processes could be active. It is, however, difficult to see how a satisfactory theory of coherent quantum behaviour in the presence of huge numbers of particles can be found. The wavefunction would be forever collapsing. We do not understand the interface between the quantum and classical regimes, but to relate our ignorance here to our ignorance of the origin of consciousness is not helpful.

It seems to me that a mix of **C** and **D** is the only one that makes sense. No amount of classical or quantum mechanical description is going to describe awareness, much less sensations and (non-supernatural) spiritual attributes. Consciousness, like the existence of Planck's constant, must, I think, be accepted as an irreducible element of the world. Brains exist, and consciousness is one of their properties. Both consciousness and Planck's constant are incomprehensible, both are real, both are properties of matter, the one of animal life, the other of inanimate matter. Whether consciousness can be engineered to exist in some structure or not will depend on the future acquisition of an understanding of the mind–matter interface that is far more sophisticated than we have now. Indeed, does the idea of consciousness being engineered make any sense at all?

Science has just emerged from stone-age billiard-ball theories; it has grasped the idea of pervasive fields, albeit with local properties; it is now grappling with the concept of a non-local, quantum field and how this can possibly be connected with space-time curvature. More germainly, it is now grasping the fact that quite new ideas are needed to understand complex systems, ideas that recognize the phenomenon of emergence and the existence of a hierarchy of levels, each demanding a new language to describe its properties. Mind is surely such an emergent phenomenon, but to attempt a description in Stone-age Speak – binary digits being equivalent to grunt and not-grunt – is somehow sad.

Some might say that we should put our faith in the science of psychology. It is, at least, a good deal removed from simplistic reductionism, though its relative youth has yet to inspire confidence. Its adherents often claim that the scientific study of behaviour is still in a pre-Newtonian state, implying that future work will reveal quantitative laws on a par with gravitational theory. Given time, the implication goes, psychology will become a 'hard' science like physics. Mental experience will ultimately be understood in terms of universally applicable laws.

This often-encountered supposition is very shaky. Psychological phenomena have a high degree of uniqueness about them because a human being is a unique individual. If more than common sense were needed to accept this, there is the authoritative claim of biochemistry that we are all genetically unique. Science can deal only with events that are repeatable, so insofar as psychology has to do with uniqueness of personality it cannot be scientific. But even when it limits itself to repeatable phenomena, as it must to be a science, its investigations will be carried out in a world far different from the physical one, a world in which the magnitude of the quantity analogous to Planck's constant will be enormous, a world in which the act of measurement may change profoundly what is being measured. In such a world Newtonian determinism is not remotely applicable and any quantitative account will have to be highly statistical in character. Mental phenomena are different from mechanical phenomena. Chalk, in short, is not cheese. Any analogy with physics is bound to be logically uncomfortable. If psychology is content to exclude unique individuality from its study, and to accept probabilistic laws to describe the rest, then it doubtless qualifies as a science. Some psychologists may think the price is too high – perhaps even a case of throwing out the baby with the bath water – but then they have to run the risk of attracting the description 'pseudo-scientists'. Prudently, most avoid dealing with conscious mental life and stick to more mechanical attributes, such as response time. They work within view **C**. But what they cannot do convincingly is to excuse their subject's difficulties by invoking analogies with pre-Newtonian physics. Whatever they are, minds are not billiard-balls.

Wittgenstein once made the heartfelt comment:

What a lot of things a man must do in order for us to say he *thinks*.

For a list of those things I can do no better than to quote P. M. S. Hacker (in his crisp account of Wittgenstein's philosophy):

> If it [a machine] can think, it can also reflect, ponder, reconsider (and there is no such thing as reconsidering mechanically). It must make sense to say of it that it is pensive, contemplative or rapt in thought. It must be capable of acting thoughtlessly as well as thoughtfully, of thinking before it acts as well as acting before it thinks. If it can think, it can have opinions, be opinionated, credulous or incredulous, open-minded or bigoted, have good or poor judgement, be hesitant, tentative or decisive, shrewd, prudent or rash and hasty in judgement. And this battery of capacities and dispositions must itself be embedded in a much wider skein. For these predicates in turn can be applied only to a creature who can manifest such capacities in behaviour, in speech, action and reaction in the circumstances of life.[10]

So if that is what it means to say 'this machine thinks', it is also what it means to say 'this machine is alive' and, what is more, 'this machine is alive like a human being'. Otherwise, the whole idea of a machine thinking is as nonsensical as the idea of the number 3 having a colour (to use another of Wittgenstein's images).

You would think after that enumeration of mental attributes that that would be definitive. Not a bit of it. Some who work in the field of AI would argue that all that is junk folk-psychology which has burbled on for millennia, and to focus a science on that is plain silly, and bound to lead to vacuous results. When neuroscience is fully worked out we will be able to dispense with those vague descriptions such as anxious, feeling good, being in love, terrified, and so on, and replace them with good precise descriptions of neural states. And, incidentally, there is no need to assume a one-to-one correspondence between a given neural state and a folk-feeling, though some certainly do. In other words, it is not a matter of replacing a verbal description which reflects a feeling of awe, say, with such-and-such collection of neurons firing; it is more a matter of abandoning the whole language of mentality, as we know it, as scientifically meaningless.

Even if this programme is viable – and why should it not be? – the resulting description of mind will not be remotely interesting to humanity at large. A science of mind is already limited in any case, in that it has to focus on mental elements that are common to all, and it has to leave out all unique experiences and many with a strong subjective element in them, so its cultural appeal and influence are going to be limited. If it also distorts the whole topic out of recognition in order to apply the language of physical science, it spectacularly misses the point. Neuroscience is fine. It is never going to be an interesting science of the mind.

The field of AI is even further removed than is neuroscience, not dealing with brain stuff, but with silicon plus a few other inorganic materials. We at least know that the immensely complex physical neuronic structure we call the brain has mental attributes. We simply do not know whether anything else is capable of mentality. To claim the possibility for inorganic machines is all right for Isaac Asimov, but curious, to say the least, for scientists when there exists no well-attested instance. There is, after all, the human body with its complex chemistry that can and does affect all sorts of mental states – mood, mental energy, emotional response, etc. Computers simply do not have bodies. Computers do not even crawl around like the Lambton Worm, picking up bits of news, not even Tyneside ones. We have to say to the strong-AI computer, 'Don't just sit there. Get a life!'

Nevertheless, there is little doubt in my mind that computing science will continue to make huge advances, and that the behaviour of machines and their interaction with human beings will get more and more sophisticated. A breakthrough may come when a certain degree of computing complexity contains within it, unprogrammed, the necessary instability for a phase change to occur and produce entirely new, maybe unpredicted, emergent phenomena. But even if these phenomena have elements suggesting analogies to human thought, they will still be machine phenomena and should be described, as they are in the rest of physical science, with non-animistic language. In its description of the physical world science decided long ago not to use metaphor and to steer clear of vitalism. If computer jargon did the same, the question 'Can a machine think?' would not arise, and the AI community could then get on with its fascinating programmes of research without confusing the rest of us.

Actually, that question simply does not arise in some views of strong AI. Consciousness is seen to be absolutely irrelevant to the performance of intelligent tasks. That humans possess consciousness is an epiphenomenal oddity that has no bearing on their extraordinarily versatile abilities. These abilities exist because of the complex connectivity and activity of the brain, both of which go on without any help from consciousness. Asking 'Does a machine think?' is irrelevant. The machine is a computationally functional being, like the brain. So, functionality is where it is at. Mental states are simply computer functional states causally related to the physical states of the brain. David Chalmers bites the bullet and goes the whole way. If computer functional states allow of consciousness, why not any functional state? Any machine with a function is therefore conscious to some degree. Thermostats are conscious beings; so are cars. Well, of course, we all knew, deep down, that cars had personalities of their own. Why else would we coax a cold engine with encouraging words, or give a car a name, or react with exasperation at its thoughtlessness in getting a flat tyre? We all knew that those machines in the kitchen can be benign or downright malevolent; boilers, temperamental; CD players, unreliable. It is comforting to know that there is now theoretical support for this widely held view of reality. But what is striking is the apparent conceptual return of some science to the magical world of three centuries ago. But not really. Chalmers's remarkable view is more in the nature of a devaluation of mentality in the strong-AI tradition than it is a celebration of panpsychism. The existence of consciousness is unimportant; it can be buried beneath the bonnet of a car for all the scientific significance it has.

Even so, the fact of the existence of consciousness will not go away, though one gets the impression that some scientists of the mind would like it to disappear. Focusing on the time-constants of the brain – reaction time, memory, etc. – happily bypasses conscious activity. Even when the existence of consciousness is taken seriously as a Darwinian survival advantage, the attempt at a science of consciousness always comes down to neurology. The experience of colour that we have is related to the spectrum of the sun and comes about via the activation of certain cells followed by the firing of certain neurons, and so on for all the senses. Even if it were possible to identify and describe the, no doubt complex, pattern of neuronal and neurochemical processes associated with my experience of listening to Sibelius's sixth symphony, the language of the description, necessarily scientific, would not come close to capturing the conscious experience. (It is difficult enough in any language.) Apart from anything else, it is simply at the wrong hierarchical level. It is somewhat analogous to an attempt to describe the performance of a transistor in terms of quarks. No theorist of quarks, gluons and quantum chromatography would dream of describing any solid-state phenomenon in those terms, even though the world of quarks conceptually underlies and infuses all of solid-state physics. In a similar way, the brain activity underlies and infuses consciousness, but a language quite different from neurology is required to provide a satisfactory description of mental events. I have no doubt that in time science will evolve a detailed description of the brain and its processes, and it may even identify special features that allow consciousness to emerge. But if the science of mind, as distinct from the science of the brain, has any relevant meaning at all, its language cannot be the mathematical one of physics and chemistry, or the language of evolutionary biology, or of genetics. It will be the language of art and the humanities. Such a language has developed and goes on developing for precisely that purpose.

Eleven

A state without the means of some change is without the means of conservation.

Edmund Burke, **Reflections on the Revolution in France**

Physics and chemistry have identified many communities of elementary particles, the elements of the Periodic Table, and how atoms interact to form molecules. They have described the properties of solids and liquids, and the rich spectrum of chemical reactions, some of them unstable and even explosive. Chemistry has combined with biology to study the behaviour of complex organic molecules in living systems, and biology has begun to understand the properties of its new atom, the gene. Science is being stretched beyond the gene to the study of the vast complexity of animal behaviour and ultimately human society. Its influence on culture, in all senses, via the aura of its success, its ambitions and above all its technology, is immense. Is it not all, as Francis Bacon wished, for the benefit of mankind?

Western society is remarkably resilient. No matter what technology throws at it, it flourishes. There is no guarantee that this state of affairs will last. From physics has come nuclear weapons, from chemistry, nerve gas, napalm and other elements of chemical warfare, from biology, anthrax bombs and germ warfare. Given the long historical evidence of human insanity in high places (and given maybe the

Principle of Plenitude) one can only grimly expect something horrible to happen in the future.

A topical concern is global warming. There is no doubt that the ice sheets in the Arctic Ocean are thinning, that glaciers are retreating, that the weather seems to be more freakish than usual. Is this because of global industrialization, as is usually assumed, or is it because of some natural solar or terrestrial cycle? I have no idea. In geologically recent times, glaciers have advanced and retreated, average temperatures have oscillated up and down by several degrees centigrade, and wasn't there a mini ice-age only a few hundred years ago? So should we really worry? Doesn't the biosphere have resources of negative feedback always to reach stability? But the emissions of carbon dioxide and other so-called greenhouse gases that have accompanied man's technological evolution are new to the planet, so the worry is that this might result in changes in climate that may well be coped with by the biosphere but at the expense of human society as we know it.

A new cause of concern comes from the genetic engineering of crops. Genetic engineering under the name of selective breeding has been going on for ages in order to improve the quality and productivity of livestock. Nobody bothers about that, even though we consume the products, because the means adopted are natural and low-tech. For the same reason, nobody bothered about scientific experiments involving the cross-breeding of plants. If Mendel had not carried out his cross-breeding experiments on peas we would not have known that inherited characteristics were not simply a blend of parental ones, but were combinations of identifiably separate properties. Inheritance was not a continuum; it was atomic. And the atom was the gene. But genetically modified crops grown in open fields seem more terrifying because they can pollinate the natural (selectively bred) crops in the neighbourhood, with unpredictable consequences. That the whole aim of genetically modifying crops is to improve the stock and increase the yield, which may or may not (so much depends on will and organization) help to reduce starvation in the world, is not sufficiently compelling to many in the well-fed West to justify the imagined, and maybe real, risks.

The Human Genome Project, the search for and identification of all the genes that people have, sounds harmless enough, but, given the reductionist fervour of some geneticists and the general reluctance of human society to accommodate diversity, it may prove to be pernicious. For a start, the Project cannot possibly investigate the chromosome structure of everybody on earth. There is a healthy diversity of genetic structure that makes each of us unique, so whatever the results of the Project, they will refer to the comparatively few involved in the research. The genetic map that will emerge will be some committee-like average of the Human Genome that will be hailed by enthusiasts as the essence of what it is to be human. Apart from obvious scientism of this sort, there is the likelihood of this Committee Genome setting a standard, with deviations regarded as unfortunate at best, inhuman at worst. One can imagine, in the worst Orwellian nightmare scenarios, parliaments in the democracies, juntas in militaristic states, their civic equivalent in totalitarian states and dictators elsewhere, promulgating new laws based on the genetic What-it-is-like-to-be-Human. The Human Genome Project is a splendidly ambitious endeavour, well worth supporting, but there is always its downside to watch out for.

A more bizarre Doomsday angst is triggered by those high-energy (and to some, those too highly energetic) physicists messing about with their ion colliders. The thing about ion colliders is that they use heavy ions, such as the ions of gold. The Relativistic Heavy Ion Collider (RHIC) recently completed at the Brookhaven National Laboratory in the States is expected to produce a plasma of quarks and gluons. There are three worries: black holes might be generated that would eat up ordinary matter; the vacuum, which far from being empty is full of quantum structure and may be only meta-stable, may make a transition to a stable state and produce havoc; 'up' and 'down' quarks may interact and change to strange quarks, a clump of which is known as a strangelet, which, if negatively charged, may be captured by a nucleus and convert it into a negatively charged object, with obvious disastrous consequences for the existence of atoms. The authoritative view, rationally justified, is that none of these will happen. But when has reason done anything for neurosis? Surely, with all this technology a catastrophe is bound to happen some time or other.

On the other hand, physics has given us the silicon chip, chemistry, plastic, and biology, antibiotics. We cannot transport ourselves instantaneously and magically to other places, but optoelectronic systems have given us virtually instant telecommunication. New materials, far beyond the dreams of the alchemists, have allowed us to fly in comfort. New medicines have alleviated physical suffering without the need to refer to the stars. Nature provides the bad and the good with a sardonic smile, knowing that we are human, all too human.

We are a mixture of bold and timid. The bold have no problem with the power science gives. Foul-ups occur, but they can be corrected; if anything is technically possible, we should have it. The timid are terrified. They wish for a safe world, a world of sweetness and light, for themselves, their children and their pets. All scientifically motivated change is dangerous, because who knows where it might lead? They wish to live in a fair, just and caring society, as far away as possible from Tennyson's nature – red in tooth and claw. They dream of Utopia. But the Utopias that the humanists have thought up in the past, even if they are not just take-offs of contemporary society, are somehow not satisfactory in this day and age. So what about science? If science wanted to do something really useful, it could design Utopia.

This is a pretty tall order for science. Society is not billiard-ball-like, nor can science say anything about value, so the chances of coming up with anything meaningful are not high. Nevertheless, let us imagine how science would go about it, casting all sensibility aside.

A Utopia is someone's concept of a perfect state. Science's first thought would be that all of those advanced in the past suffer from the fundamental flaw of presupposing permanence. But nothing is permanent. Stars evolve, and, as the Neoplatonists might have put it, as above, so below. The characteristic time-constant is – in round numbers – 10^{10} years for a star, 10^8 years for a continent, 10^6 years for man, 10^4 years for a civilization, 10^2 years for an individual, 50 years for economies, 5 years for business, 1 year for crops. Even in the world of matter, thermodynamic equilibrium is an ideal never attained. The continuous interactions that occur make the achievement of a steady state, in which forces balance, a condition maintainable only for a limited period. The same is true of biological systems. Indeed, periodic behaviour, a cycle of growth and decay, is more characteristic than long-term stability. Why have we not been hitherto entertained by the vision of a Utopia in which periodicity is fundamental? But then even a fixed periodicity will not do. In real dynamic systems periods often change, sometimes becoming longer, sometimes shorter, and sometimes the system becomes chaotic and in many cases the evolutionary path is quite unpredictable. Society is, among other things, a real dynamic system and as prone as any other to evolution and even chaos. In short, no Utopia is going to possess permanence; in any vision of society, change has to be incorporated.

But the idea of change is anathema to the system-builder, and it frightens a lot of people. To avoid it one has to pay a price. Plato and the Eleatics along with much of Eastern religion saw change as illusory. Reality lies with the Eternal Forms, the God that is everlasting. What is important is the individual's path to salvation, to Nirvana, to the union with God. In Jewish and Christian mysticism the communion of the individual with God is central, with ten rules of behaviour carved out eternally on tablets of stone and, for Christians, an eternal exhortation to a new kind of love. Islam's rules are also for eternity. For all religions, change is irrelevant – the Truth was revealed many years ago. Social institutions can be what they may, rendering to Caesar what is Caesar's, but what matters is the individual soul and its salvation in the cosmic battlefield where good and evil struggle eternally. Regarding change as fundamentally illusory implies that if a Utopia is contemplated at all, it cannot be of this world, but of the next, in the form of the Kingdom of Heaven.

Would-be social reformers can have no truck with this sort of thing, even though they recognize the human need for a meaning to existence. To the classical Marxist, still trapped in eighteenth-century materialism, such a need is a product of the wrong sort of society. Religion, the opium of the people, will not be necessary come the communist millennium; individuals, completely determined by the society they live in, will lose the need for it. Change is inevitable but not a problem once the state has withered away, since society and its members, obeying immutable deterministic laws, will be in perfect harmony. Such a vision is difficult to sustain in the light of twentieth-century science, under which, if it is still possible to be a materialist, it is no longer possible to be a determinist. The quantum nature of matter manifests a world that is indeterministic at the most fundamental level. Even in the world of classical physics there are dynamic systems whose behaviour defies accurate predictions. And if this is true of Newtonian mechanics, how much more serious is the situation for a system as complicated as human society.

Politicians, businessmen and bold spirits in general will face the impossibility of achieving certainty in any activity with equanimity. What actions are ever carried out on the basis of full and perfect knowledge, after all? But we must consider the problem of a Utopia that addresses the needs of those who are haunted by images of justice, freedom, equality, democracy, greatness and everything whose lack leads to a feeling of dissatisfaction with the present society and the yearning for better. Understandably, perhaps, it is the socialist or social-democrat who is most likely to be moved by the thought of ideal institutions. Whether based on dissatisfaction with present society, or on the fear of any society not closely regulated, the propensity of the left, from Marx onwards, to advance ideal, occasionally labelled scientific, solutions is probably inexhaustible. Such solutions are typically characterized by adjectives such as logical, rational, correct and justified as following from self-evident premises, such premises leading inevitably to the subordination of the individual to the state. And, needless to say, there will be no admission of change within, or any analysis of stability of, their ideal system.

Yet the left does not have the monopoly of Utopias, though the ideal institutions of the right are very different. Whereas left-wing Utopias are immanent, as it were, those of the right are transcendental. The former are basically materialistic, informed at best by the perceived needs of the masses for welfare, and judicial and social equality; the latter are quasi-religious, uplifted by concepts of national or racial superiority and addressing the perceived needs of an elite. The right is plainly less rational than the left. It draws its strength from deep-seated feelings about hearth and homeland, from a respect for strength and excellence, and an awareness of the need for defence against a potentially hostile world. The general intellectual principles that motivate the left and persuade it that international scientific laws of society exist appear utterly foreign to the mind of a right-winger, more at home with his intuition about the existence of a 'natural order of things'.

The crude polarity of left and right that I have sketched out reminds one of Nietzsche's slave morality and the morality of the nobles in his *Genealogy of Morals*, published in 1887. Nietzsche describes a fundamental dichotomy in social morals and relates its origin to military conquest. When a nation conquered another, the leaders of the victorious nation became nobles and the mass of the subjugated nation became slaves. The nobles practised a morality amongst themselves that emphasized honour, courage, physical prowess and contempt for slaves. The slaves, full of resentment and living miserable lives, had to invent a power greater than the nobles and a morality that emphasized the virtues of mutual aid, humility and suffering, in order to salvage their self-respect. In effect, the slaves invented their own religion, defining bad as what nobles did, and dreamt of a just and fair society and rewards in the hereafter. Nobles who happened to scorn aristocratic life, perhaps because of their frail constitution, saw an opportunity for a different sort of power and latched on to the slaves' resentment, becoming priests. Nietzsche's account of the origin of morals is a far cry from Christianity and the ethics of Plato and Kant. Crude though the division is, we can recognize the outlines of master morality and slave morality, both alive and well, in our own society.

Change within a society, whether brought about by science and technology or by climatic or cultural shifts, is a serious matter for all institutions, but perhaps more so for those on the left. Classical materialism cannot survive indeterminism. The health service of the welfare state cannot maintain its ideals in the face of ever more ingenious but expensive innovations in medical technology. The welfare state itself finds difficulty in coping with rising expectations. For the right, it is very much the case of plus ça change, plus c'est la même chose. Nothing can undermine the old-fashioned conservative values in the long run; rather they are likely to be reinforced. Fashionable tinkering with society cannot alter centuries of human behaviour; technical innovation is to be exploited, not feared. Social change could, of course, be uncomfortable for the right, but not a fundamental threat – the elite will continue to govern, if as party leaders in the communist state and not as gentry. If change is to be resisted, at least temporarily, neither left nor right has any choice but to plump for authoritarianism, whose ultimate expression is the totalitarian state.

Convenient though the polar concept of left and right is, it does conceal the fractal nature of politics. Zoom in to any activity of any political party and one will find divisions of opinion which attract the labels of left and right, and with as much animus as one would find between parties. And there may be factions within factions. Where divisions are intellectual in nature the fractal property is likely to be well developed and well defined; less so where divisions are emotive. On this basis the left is likely to be more fractal than the right. A fractal number can be envisaged that measures the degree of party disunity; zero for an utterly united party, 1 for a party riven by well-defined factions, 2 if even the factions are significantly subdivided, and so on. But this is a diversion.

We can now say something useful about a Utopia. If change is to be feared, then some form of totalitarian state is desirable. Whether it is to be of the left or right is probably unimportant. But it should be said that such a Utopia would be dead or, rather, in suspended animation. Change will come willy-nilly, and being pent-up it may burst upon the society with destructive force. Really, I think that we, as scientists, cannot recommend this option. But that means that we have to consider a novel sort of Utopia, namely, a time-dependent Utopia.

This naturally kills the traditional concept of a Utopia straight away. Unfortunately, the idea of a time-dependent Utopia misses the whole point, namely, that Utopias are ideal societies, so what has reality to do with it? If things in nature are time-dependent – so much the worse for them. There are only two things that Utopia can be, *ou topos* (no place) or *eu topos* (nice place). The latter is the one wanted.

But surely a possible nice place is better than an impossible one. Surely it is interesting to see how close to a nice place one can really get. Let us suppose that, grudgingly, science is allowed to go on.

The time-dependence of a society may be considered to fall into two categories depending on whether the change is continuous or abrupt. Into the latter fall violent acts such as coup, rebellion, war and natural disaster. There may be Utopists who feel that an ideal society should include such events, but we do not really believe it, though certainly no society can ignore the possibility that such events may occur. Acceptable, time-dependent Utopias are therefore those in which change is relatively slow.

Change is fuelled by two sorts of innovation – conceptual and technological – and there can be long-term (secular), short-term and periodic aspects. Conceptual innovation can mean new and evolving ways of appraising man and mankind that leads to secular changes. These arise from religion, philosophy, science and the study of society, and it is this that, in crystallized form, informs conventional Utopias. The way in which man views himself and his world naturally affects the form and function of his society. God or no God, man as spirit or man as matter, fixed hierarchy or mobile meritocracy, the ideas of progress, justice, equality, democracy, morality, the Rights of Man, expectations of standard of living – any change in such ideas constitutes a powerful dynamic entity, generally acting over more than one generation. Conceptual innovation can also be less elevated and more focused on current topics, such as a change of view regarding the efficacy or desirability or popularity of the party in power, which in a democracy can lead to change of government. Technical innovation is equally powerful. Navigation, gunpowder, steam power, radio, motor cars, aircraft, nuclear energy, computers, space travel, genetic engineering – every new process, every new device, increases our ability to manipulate the physical and biological world, presenting new freedoms and new problems. One consequence of the global nature of technical advances – transport, telecommunications – is that severe limits are now set to the degree of freedom to create a Utopia that is not world-wide. Attempts to inhibit conceptual change may occur in totalitarian states, and in the softer democracies there is a tendency to wish away technological change, but both are sure to be utterly vain in the long run, given the vitality of man's intellect and ingenuity, which are forces as real as gravity and electromagnetism, and just as intrinsic to the cosmos.

Attempts to incorporate all these dynamic aspects consequent on conceptual and technological innovation in a coherent and planned Utopia seem doomed to failure. Nevertheless, one might start by guessing that whatever the details of such innovations, what matters institutionally is the form of government and response of the people to the way the government handles change. The dynamic that matters here is public opinion, and all the time-dependence may be seen to come about through changes in public opinion. But this can be effective only in some sort of democracy, so we will henceforth suppose that our Utopia is democratic. We may also guess that public opinion, though by no means changing smoothly with time, exhibits an inexorable trend from supporting a government at some epoch to opposing it at a later epoch. A youthful, adaptable and ingenious government may retain public support for many years, but ultimately – such is the way of all living organisms – the vigour diminishes, the ideals are less bright, the blood runs colder, and public opinion will turn against it. Opposition eventually follows support. The image of perpetual support is too incredible, though all Utopias hitherto contain such an implication. So our admission of a degree of cussedness vested in the people is a decisive demarcation between the old and new Utopia.

The dynamic of public opinion leads to periodic change, though with no fixed frequency. In a first-past-the-post democracy, R, a rightish government, is elected with modest public support. This may initially wax, but eventually it will wane. Eventually an election transfers power to L, a leftish government, and the pattern of support repeats itself. Time is carved up into distinct periods in which positively rightish or positively leftish policies are pursued which encourages positive opinions and positive changes of opinion. Things are different where there are attempts to achieve a closer correspondence with some ideal of democracy. With proportional representation one tends to end up with a coalition after every election with inevitable blurring of action. Under these conditions it is impossible for opinion to be either positive or effective. This has the important consequence of eliminating change, or at least minimizing change, at elections. Unfortunately we have to guess that the amount of change to be expected at an election and voting apathy are complementary variables in the sense that small change engenders great apathy and vice versa. The gain of small change might have to be paid for in the ultimate demise of democracy through voting apathy, i.e. proportional representation runs the risk of killing off democracy.

The Burkean ideal of democracy is government by elected representatives and not, as becomes increasingly common, elected delegates. But maybe the literal idea of democracy carries within it the cancer of its own destruction. Consider the technological innovation that allows each of us to vote on every issue, registering our votes electronically. Is this not true democracy? But governing is a continuous activity, which implies, in an extreme interpretation, that we could do nothing but govern, poised with our fingers over the voting buttons while we weighed the pros and cons. But if we all did that, there would be no other activity to govern. What price democracy then? But, of course, this is nonsense. After the novelty had worn off people would ignore their voting buttons and get on with their lives. Only the politically ambitious or politically motivated would continue to be involved, but then they always have been involved, so there would be no change in that respect.

So where does that leave us with the problem of Utopia? We have seen that whatever institutions we prescribe must be adaptable to conceptual and technological innovations that are virtually impossible to predict. No rigid planning is possible. Some sort of democracy seems to be better than totalitarianism, but it probably has to be world-wide. There remains only the problem of stability against change. We have already noticed that the rigid regimes of totalitarian states are ultimately unstable. One reason for this, even in benign regimes, is that the more structured and rigid a society is, the more is change noticed. In a non-rigid, free society there is nothing for change to topple. The analogy in physics is a system of maximal random motion, as occurs at thermo-dynamic equilibrium, a system of maximum disorder. Change affects order: the less order to begin with, the less effect. The moral here is that our Utopia should be one in which every conceivable human activity consonant with minimum social order is encouraged, whether seen to be socially useful or not.

This Utopia is, unhappily and not surprisingly, a far cry from a vision motivated by the need for a safe, just and caring society. It looks suspiciously like a Western democracy, but emphasizing adaptability to change rather than welfare, and almost certainly with a laissez-faire economy to maximize stability rather than a planned one. It need not be unsafe, unjust or uncaring, but it must first make sure that it is vibrantly alive and not a sickly organism.

And that is as far as science, or at least as far as this scientist, cares to go. So much for the concept of Utopia, a nice place. The no doubt penetrating analysis given above naturally misses the point. The yearning is for myth, nothing less. Reality, the world of earnings and debts, of illness, of injustice, of efficiency, of management, of change, should not intrude. In matters of this nature science naturally gets it wrong. We said as much at the beginning, and we were right. Science is the last discipline to consult about ideal societies.

Science and Art
Twelve

Even if art is, among other things, an enquiry into the nature of our feelings about things, it is an enquiry that seems, prima facie, to be intrinsically different from the general enquiry into nature that is science. In some ways, art and science can be seen as opposite sides of the same coin, the one illuminating self-knowledge, the other public knowledge. Knowledge about the natural world, of which we and our emotions are a part, connects the two, even though the truths revealed are very different animals. But there is a utilitarian flavour in this sort of connection that need not be insisted on; equally there is the excitement and gaiety of creating something of special significance, be it a painting or a research paper – both are painstaking, absorbing activities which offer that intense, characteristic satisfaction that creativity has. In this respect laboratories and studios have a lot in common.

But in stressing what may be common, we must really avoid the pathetic attempts that are often made by scientists to engage the attention of the artist by pointing to pretty pictures of nature generated, for example, by electron microscopes, or to the Laura Ashley patterns of fractal mathematics. 'Look', they seem to say, 'science is not that dull. It is even artistic.' This sort of thing can only bemuse the artist and reinforce his opinion that scientists are a philistine lot. Pictures generated by microscopes or mathematics are not art. That they have been put forward as such is at least evidence that they have evoked some sort of aesthetic response in the scientist. But that does not make them art. Absolutely anything in the world can evoke an aesthetic response. Personally, I have found the printed circuitry in my computer quite beautiful to look at. And who has not found the odd bit of curiously shaped wood evocative? An aesthetic response is one thing; art is another.

Not everyone would agree. Surely the printed circuit, at any rate, is a product of art of some sort. Of course, that is true enough, but the art of the artisan is not art at all: it is craft. R. G. Collingwood has distinguished the two perceptively.[1] The basic difference between art and craft is that craft knows where it is going and art does not. When a craftsman starts to make something, he knows what that something is and knows when he has made it. The artist, though he may have a plan and general intention, inevitably employing craft galore, knows what he has created only when he has finished. There is this element of discovery in art that is missing in craft.

In this respect, art is similar to science. Art aims at discovery via the manipulation of its chosen medium – paint, stone, words, sounds. The medium of science is more abstract – experimentation, mathematics – but its aim is discovery. Like art, it does not know precisely where it is going. Loads of craft go into the design and execution of an experiment. The intention is to explore further some phenomenon to reveal something new or to deepen the understanding and knowledge that already exist. Mathematical craft can be exploited to create elegant summaries of experimental data and to discover new theoretical structures.

Craft and technique are always present in art and in science, but in many cases there is little else. Hack science as well as hack art is not unknown. There is a continuous spectrum stretching from low-hack to high-art. Bodies set up to fund science naturally do not want to fund more-of-the-same, but nor do they dare fund projects that are too way out. Science builds on established foundations. In this they appear to differ from corresponding bodies supporting the arts, at least in the UK, which, afflicted with a sort of scientism, appear to fund and award prizes to stuff only if its prime virtue appears to be novelty. The motto is if it hasn't been done before and the candidate says it's art, it must be good. But art, like science, must build on firm foundations. What has been found beautiful in the past and is still regarded as beautiful may and perhaps should inform creativity in the present. Much of modern architecture and modern art in general has abandoned that idea in orgies of individualism and quests for novelty. Fortunately, the essence of scientific practice makes perversion of that sort virtually impossible.

Like science, art had to free itself from religion in order to develop. The history of art is as fascinating as that of science. The discovery of perspective and the techniques for depicting it, the movement away from icons to real people, real scenes, real landscape, parallel science escaping from astrology into astronomy. Like science, art discovered nature. Bellini's Madonna is a girl we might meet; Carpaccio's crowds are full of people we might know; Dutch genre sometimes gives us people we may not want to know; Vermeer is Jane Austen in paint. The smudgy landscapes hiding behind portraits like Leonardo's *Mona Lisa* burst forward in the works of Corot,

Claude and Constable. Up to the nineteenth century it could be said that both science and art were exploring nature in their different ways, but then art got bored, or perhaps it anticipated the fuzziness of quantum mechanics, and settled for giving impressions of nature, and moving to the expression of feelings. Maybe the growth and power of science forced art to concentrate on its more subjective elements in order to emphasize the difference between science and art.

One of the characteristics of art that tends to make it a hostage to relativistic fortune of the 'anything goes' type is that it seems to defy definition much more than does science. There is no necessary and – more to the point – sufficient definition of what constitutes a piece of art. We believe we know when we see it, but we cannot define it comprehensively. Science is fundamentally no different in this respect, though its permeating rationality suggests that it ought to be, but it is far less prone to 'anything goes' relativism (except in the minds of some sociologists of science). Beyond a certain point it is simply uninteresting, and inevitably unproductive, to search for tight, comprehensive definitions of any human activity, art and science included.

One difficulty in this respect is that whatever art is about cannot be separated from its medium. The colours of the Italian Renaissance, the brush strokes of the French Impressionists, the abstracts of modern art, point to the fact that painting is as much about paint as it is about what is expressed. In sculpting, a figure in marble is not the same as it would be in bronze. In poetry, the words themselves are as vital as what they mean. Music provides the most compelling examples: compare Beethoven's Grosse Fuge quartet with the version for string orchestra, or the string-orchestral version of Schönberg's *Verklärte Nacht* with its adaptation for piano trio. The medium itself is part of the art.

Surely, this is not the case in science. But is that true? Can science be as much about experimental method and mathematics as about nature? There are certainly different styles of science. In my own field the sheer complexity of the trillions of physical interactions that go on between the particles of a cubic millimetre of a solid make it necessary for any handleable mathematical description to rest heavily on approximations. Carrying the art (or rather, craft!) of approximation to the limit may allow one to obtain an analytic description of a whole range of phenomena in the form of a more-or-less simple mathematical equation. Alternatively, one can opt for far fewer approximations and describe a particular case numerically using a computer. Number crunching is more accurate, in principle, but its results refer only to a specific situation. Analytic methods provide general descriptions, but are less precise. The ideal is to do both, but it is perhaps surprising how few examples there are of that. It seems that physicists are temperamentally inclined to one approach rather than another. However that may be, the result cannot be divorced from the way it was obtained. The science seems to be really about number crunching or analysis as much as it is about nature.

Nowhere is the choice of experimental method more significant than in the investigation of quantum phenomena. An experimental set-up designed to measure particle-like properties of a system will yield particle-like results; one designed to measure wave-like properties of the same system will yield wave-like results. Here the medium – the experimental method – is as important as the result. In fundamental theory the rôle of the medium – the theoretical approach – is equally important. Taking the velocity of light to be a fundamental constant of nature leads to the conversion of gravity into curved space-time. Keeping space flat and Euclidean restores gravity but lets gravity alter the speed of light. Our picture of nature depends on the medium. Are particles point-like or string-like? Are there more than three dimensions of space and one of time? Can the choice of imaginary time solve the problem of the Big Bang?

At these frontiers of science the medium and its manipulation are central, and, indeed, constitute actual science in this area. One begins to see the rise of schools, each promulgating its own line, its own interpretation of the meaning of science at the frontier. In quantum theory the Causal Interpretation is essentially an evocation of the classical form, the mathematico-logical interpretation of the abstract form. It may be stretching a point to see the Copenhagen interpretation as impressionism, and as for the Many Worlds idea – who knows? In the business of going back in time to the first femtosecond or so after the Big Bang there is a sense of mysticism that stimulates the creation of wonderful theories, pointillism giving way to stringillism, as it were. Perspectives are no longer 3D, but rather multidimensional. Pictures of nature change accordingly.

So there is a case to be made that science is no different from art insofar as the medium occupies a defining rôle. But while that may be the case, it is what science and art say, however medium-laden, that is ultimately of interest. Science purports to tell us about the physical world, and is demonstrably successful, as technology testifies. What art is about is less clear. The persuasive view of Collingwood and others is that art includes the expression of emotion and at the same time the *defining* of emotion as distinct from the *betrayal* of emotion. But emotion is not all that art is about. There is the perception of the world, and this can be largely dispassionate. The precise nature of the emotion or perception becomes clear only on the completion of the work, and in this way art is a source of self-knowledge. The discovery is made when the painting or poem is finished. Sometimes there may be the nagging feeling in the artist that there is more to discover, that something is not quite there. Cézanne's fascination with Mont Saint-Victoire suggests something of the sort. But again, there are parallels in science. Until a research paper is written, the scientific discovery is not precisely formulated. The writing up is as much a part of the science as the research on which it is based.

It seems that there are many aspects of art and science that are shared, perhaps not surprisingly in that both further knowledge, the one, knowledge about one's emotive and aesthetic responses, the other, knowledge about things outside one's self. But there is one crucial aspect in which they differ. The work of art is a unique object, untranslatable. It is that it is. It is a celebration of the human condition. There are many works of art and many sources of aesthetic experience, each of them unique, and it is this plurality that allows comparisons and evaluations to be made, in short, the articulation of the aesthetic experience in an objective mould. The work of science is not, other than in the trivial sense that this particular scientist (or more likely these days, this particular team of scientists) actually did it. It could have been done by somebody else and, no doubt, the presentation and emphasis would have been different, but what constitutes the science has the character of universality, with an essence that transcends different formulations. Change a word in a poem and it becomes a different poem. Add a brush stroke and the painting becomes a different painting. This is the nature of art. It is not the nature of science.

The fact of the matter is that the world, with elements and events that are unique, is intrinsically more like art than science. That uniqueness is most evident in ourselves, and we rely on art to tell us something about our unique complex nature. When we look outside ourselves we are bewildered by the infinite variety of the world and we cope with it only by mentally reducing that complexity to simple repeatable events, rationally described in the way science describes things. In this way art and science complement one another.

It should not be forgotten that aesthetic principles suffuse science itself. The rational description of the natural world is carried out with as much elegance, economy and simplicity as possible. Fractal patterns, suitably coloured on a computer screen, may be pretty to look at, but the elegance is in the simple mathematical equation that generates them. A multitude of electrical and magnetic phenomena is encapsulated by just four equations discovered by Maxwell, and these can be reduced to a single equation in special relativity. The power residing in that single equation is awesomely beautiful. If the essence of science is not aesthetics, as it is for art, science, nevertheless, cannot be indifferent to it. The similarities between art and science are fascinating, but so are the differences.

Science and Sensibility
Thirteen

The finest discoveries concerning culture are made by the individual
man within himself when he finds two heterogeneous powers
ruling there. Supposing someone is as much in love with the plastic
arts or music as he is enraptured by the spirit of science and he
regards it as impossible to resolve this contradiction by annihilating
the one and giving the other free rein, the only thing for him to do
is to turn himself into so large a hall of culture that both powers
can be accommodated within it, even if at opposite ends, while
between them there reside mediating powers with the strength
and authority to settle any contention that might break out.

Nietzsche, **Human, All Too Human**

The state of man would be indeed forlorn
If false conclusions of the reasoning power
Made the eye blind, and closed the passages
Through which the ear converses with the heart.

Wordsworth, 'The Voice of the Universe'

It is time to sum up. This book has been about science, something of its history, its escape from magic, its relationship
with mathematics, and its tendency to evoke scientism. Some
of its conclusions may be summarized in the following
points.

1. Science has fundamental limitations. Among the most
important is its inability to say interesting things about
consciousness, ethics, art and religious belief. The belief

that science can say interesting things about these topics is scientism. It is a belief founded on a monumental category error and utterly misguided.

2. Natural magic, magic free of its supernatural and superstitious baggage, is still a meaningful part of the human world. It stimulates the imagination in a way that is unique, and it informs religion, art and even science in terms of real forces that 'move' people.

3. The divide between the Two Cultures, Science and Literature, is bridgeable only by recognizing the basic complementarity of science on the one hand and art on the other. Science, unlike art, deals solely in public knowledge; art, unlike science, deals in self-knowledge.

Examples of weak, moderate or strong scientism that I know of are, revealingly, all within science. The accompanying presumption is the meaninglessness of whatever science cannot tackle. It would be odd to find such a view expressed by a scholar outside science, at least in our modern age. Yet it might be a very common belief among the vast majority of people if they thought about it, impressed as they are likely to be by the power of science. It may even be held in a somewhat unthinking way by some intellectuals. Science is extremely powerful. It is also foreign to most. Most people can understand and enjoy and even analyse painting, music, sculpture, drama, literature, without specialized knowledge. Such things are immediately accessible, whereas science is not. Science is for scientists and maybe a few philosophers. For the rest, science is likely to be regarded with a kind of xenophobia.

After all, people live more in a world described by literature rather than science. Look at the following list of subjects.

Science	Literature
Elementary particles	Initiation and maturation
Cosmology	Parents and children
Structure of matter	Men and women
Chemical processes	The individual and society
Genetics	The minority experience
Botany	The artist and art
Anatomy, physiology, etc.	The order of nature
Animal behaviour	The extraordinary and the fantastic
Anthropology	Terror and violence
Psychology and social studies	Ageing, dying
Medical science	Humour

The extraordinarily different nature of science and of literature is striking. Whatever the subject of enquiry, science must apply its methods, with all their power and all their limitations. As a consequence, the results of any enquiry will refer to a population but never to an individual. Literature, on the other hand, has no formal methods, its worth limited only by the talent and insight of the author, but it speaks, if it speaks at all, of and to the individual. Psychology and social science may study art and the artist or the individual and society, but their conclusions will refer to the special group of artists or individuals that was chosen, which will not necessarily apply to the individual artist or the individual in society.

If curiosity about nature, the exercise of mathematical ingenuity, the heady feeling of 'reliable knowledge', inform the scientist, what is it in people that responds to the celebration of human life that is literature, art and the humanities?

What is it that motivates people's delight in landscape, sunsets, gardens and, more abstractly, elegance and form? I believe a useful answer is sensibility.

To define sensibility is to define what permeates the whole of the humanities, the whole of art, and, it would be splendid to think, the whole of personal relationships. It is at base a cultured, emotional response to the actions of fellow humans, an empathy for the other person, a refined feeling for and recognition of form in manners, art and nature. It cannot be coldly Apollonian or fiery Dionysian, but mediates confidently between those extremes. It belongs to a spirit that has a sense of humour and a wry acceptance of human behaviour. It would take some delight in observing that your eighteenth-century English gentleman would unhesitatingly allot the Pride to Darcy and the Prejudice to Elizabeth, whereas your twentieth-century American might well reverse that allotment. However passionate about art, it would appreciate the put-down definitions: poetry is the stuff in books that doesn't reach the margins; painting is a way of protecting flat surfaces from the weather; music is the result of reading a bar-code the wrong way round.

So wags the world. To try to define sensibility further would be to lack it, so I won't. Asking What is sensibility? is a bit like asking What is jazz?; the famous response to a lady who asked the question being, 'If you have to ask, ma'am, you'll never know.'

As a physicist with only moderate sensibility, I can't resist risking an analogy with dynamics. Sensibility is what is influenced by the forces of what I have termed natural magic. Like intelligence or any other mental attribute, sensibility shows a substantial demographic and cultural variation, but, at base, it is what causes a person to respond positively to music, art, rhetoric, etc. It is the 'charge' in the conscious mind that is responsive to the forces of natural magic. In effect, this defines what I mean by natural magic in terms of the broadly accepted meaning of sensibility. Francis Bacon defined magic as follows:

> We here understand magic in its ancient and honourable
> sense – among the Persians it stood for a sublimer wisdom,
> or a knowledge of the relations of universal nature.[1]

Subtract science from magic 'in its ancient and honourable sense' and what remains is my natural magic. Forces exist in the natural world that act through the senses on the mind and 'move' people emotionally They are as real as gravitation and they are certainly more than metaphor. They need a name, and that name is natural magic. These forces undoubtedly exist in nature, as real as anything gets.

An example of what I mean by a non-scientific force is provided by the writing of Thomas Hardy in *The Woodlanders*:

> The physiognomy of a deserted highway expresses solitude to
> a degree that is not reached by mere dales or downs, and
> bespeaks a tomb-like stillness more emphatic than that of
> glades and pools. The contrast of what is with what might be,
> probably accounts for this. To step, for instance, at the place
> under notice, from the edge of the plantation into the
> adjoining thoroughfare, and pause amid its emptiness for a
> moment, was to exchange by the act of a single stride the
> simple absence of human companionship for an incubus of
> the forlorn.

Hardy is describing the effect on the mind of being in a certain environment. The effect is real. One can imagine oneself responding somewhat as Hardy describes, but there is more to it than that. Hardy's art – word magic – would intensify an actual experience; art has its own power. Look at another example – Hardy's creation of Egdon Heath in *The Return of the Native*, a haunting example of spirit of place:

> A Saturday afternoon in November was approaching the time of twilight, and the vast tract of unenclosed wild known as Egdon Heath embrowned itself moment by moment.
> Overhead the hollow stretch of whitish cloud shutting out the sky was a tent which had the whole heath for its floor. . . .
>
> The distant rims of the world and of the firmament seemed to be a division in time no less than a division of matter. The face of the heath by its mere complexion added half an hour to evening; it could in like manner retard the dawn, sadden noon, anticipate the frowning of storms scarcely generated, and intensify the opacity of a moonless midnight to a cause of shaking and dread.
>
> In fact, precisely at this transitional point on its nightly roll into darkness the great and particular glory of the Egdon waste began, and nobody could be said to understand the heath who had not been there at such a time.

The idea of understanding a heath – in general, understanding a spirit of place – makes no sense in science, but plenty of aesthetic sense. A different reality is being described here. Different forces are in play.

The concept of force has emerged from the observation of motion – of billiard-balls in the simplest scenario. But ordinary language speaks of people being moved, in the emotional sense, and it is in this sense that non-scientific forces exist. They exist because human beings are conscious and self-aware and rationally aware of their environment. These human beings are yet composed of quarks, electrons and photons – nothing supernatural about them – but non-billiard-ball forces have their conscious effects – the movement is of an altogether higher order of natural existence. Science, in its original and deepest meaning, cannot illuminate these phenomena of consciousness using the methods and mind-set of science as we know it today. All that can be done with the scientific method in this context is to observe and measure public responses, but this provides knowledge only of a statistical nature, and only of a population.

Probabilistic knowledge of this kind is fine if it is about fruit flies, but it can be dangerous if about people. One danger lies in the common misapprehension about what statistical knowledge is. A property belonging to each member of a population, to which a number, measuring magnitude, can be attached, is identified, and the subsequent statistics, at their simplest, tell us what the average magnitude of the property is and what is the deviation from the average. The common tendency is to put values to these neutral measures – norm equals normal, deviation equals deviant – so instead of rejoicing when the results show a rich spread, there is anxious concern that people are so different – life would be neater, safer, if we were all equal. Another danger is scientism, the view that what is important is only what can be measured. Such a view spawns ideas of statistical morality, statistical aesthetics, statistical values in general, which might be operationally fine for, say, the business of justifying continuing expenditure on public broadcasting to accountants but monstrous otherwise.

Human devaluation via statistics is bad enough, but there are two other dispiriting forces that are more modern and, if anything, more potent. Ever since Alan Turing devised his test for a human-like computer that gave responses that were indistinguishable from those of a person, we have become computers. We can be switched on, we can be programmed, we can process data, we can calculate and we can crash. We are basically computers with built-in software and hard wiring that has evolved over the millennia. In future we may be downloaded on to floppy discs and stored until doomsday. All the language of computerese applies to us, and with repeated usage we come to see ourselves as computers. But at least computers are active and useful, even powerful. Modern biology, however, would reduce us to the helpless playthings of our warring genes. What we are, what our relationships are and even what our cultural activities are, are determined by what our genes want and how they plan to survive us transient individuals. They make men and women mutually attractive, enough to have sexual intercourse so that their genetic reproductions have the chance of surviving in longer-lasting young bodies. You fall in love by order of your genes; and even what you fall in love with is for the most part determined by your genes. There may even be, for example, a homosexual gene. But what about the intellectual life? You think, therefore you are? Genes again. We begin to ponder, is there a Stoic gene, an Epicurean gene? Or, to turn full circle, is there a geneticist gene? The game seems to get ridiculously bizarre. Yet genes do control the colour of our eyes, our sex, our resemblance to parents and grandparents and, more sinisterly, our proneness to particular malfunctions. We are, whether we like it or not, genetic machines, computer-like, chimpanzee-like, full of instincts and instinctive behaviour. And all the language of the AI people and the evolutionary biologists is properly applicable – no question.

So what? Just as insisting that our bodies are made of

quarks is largely irrelevant in any human context, so is the insistence that our brain is a glorified computer and that our behaviour is really determined by our genes. For a start, if the brain is in any sense a computer, it is so far in advance of anything based on silicon that the comparison is premature, to say the least. And as regards our behaviour being determined by genes, where, outside pathology, is the evidence? Genes often interact with other genes, so the effect of a single gene is often fuzzy. On top of that there are the immensely complex interactions of the body chemistry to make even the influence of a group of genes fuzzy, never mind a single one. So within the rationale of science itself, the case for the brain-computer and the case of what might be summarized as gene morality are simply not made.

But suppose these cases were made. Suppose we were all absolutely convinced that, yes, our brains are in a real sense computers, and yes, our behaviour, our morality, our religion, our aesthetics, were all determined by the particular sequences of nucleotides on the DNA of our chromosomes; suppose all of that. What then sang Plato's ghost? Or, to put it less succinctly, how would that affect happiness, delight, love, awe, reverence, determination, pity, good, evil – the quarks of humanity? I suggest that the answer is, not a jot. People have been telling us that we are just machines of one sort and another for 200 years. Of course, we are. We are apes. Of course, we are. To be precise, we are chimpanzees. So be it. What has that got to do with anything of the slightest importance? We feel ourselves to be unique individuals with a unique destiny and a unique outlook. We are that we are. But that conception of ourselves is so vulnerable to the wrong language. If people keep telling us that we are simply machines, we begin to behave like machines, and sensibility disappears. But one thing is certain, our sense of uniqueness will remain and the need to service all those spiritual quarks will continue undiminished.

The fact is that unique, individual, human beings are anathema to science. Yet unique, very individual, human beings do exist, their complex behaviour surely not beyond reason, if beyond quantification. Morality and ethics are discussed and understood rationally and their genesis in the ancient societies of humanity can be accounted for.[2] Our perception of form, vital to science, for example, our intuition of space and time, which Kant argues to be given *a priori*, may be reasonably taken to be a factor aiding survival in an indifferent world. Art and aesthetic sensibility grow naturally out of the play of the human animal with forms and with matter. The interactions between individuals, between an individual and a group, between groups . . . these dominate the conscious life of a person, inevitably stimulating intense interest in moral and aesthetic truths. We wish to understand ourselves better, but we are aware that reason is not enough, language is not enough, even knowledge by acquaintance is not enough. We need art. We need the gifted insights of the artist to reveal these truths. The gifted insights of the scientist point in an altogether different direction.

While it seems obvious that humanity needs both sorts of insight, this has not stopped the more fanatical or more frightened advocates of art on the one hand and science on the other carrying on a sniping war over the centuries. Keats, concerned by the sheer rationality of science, predicted the disappearance of poetry. Matthew Arnold advocated the 'sweetness and light' derivable from reading 'the best that has been thought and written in the world' as an antidote to the barbarism of the Industrial Revolution. F. R. Leavis in the Two

Cultures controversy of the 1960s fulminated immoderately against C. P. Snow's suggestion that the literary establishment could benefit from being acquainted with the Second Law of Thermodynamics. More recently, sociologists of science, post-modernists and cultural studies experts have sought to condemn science by suggesting that its account of nature merely reflects the dominant ideologies and power relations of the culture that produced it, and that it was therefore without objectivity.

To some degree, science has asked for it. Long ago, the idea that science would progress indefinitely and naturally extend to all spheres of human knowledge was roundly debunked. The idea, nevertheless, lives on. That science could say knowledgeable things about value – the so-called Naturalistic Fallacy – is all too readily believed today. Particle physicists search for a Theory of Everything – a nomenclature and choice of words that reveal the kind of mind-set that makes science seem antagonistic to the human spirit and scientists seem arrogantly magus-like. A recent book entitled *The Physics of Immortality* blandly presupposes that it will be possible in the future to encode an individual human being in a computer program to be reborn (down-loaded) whenever.[3] The possibility of identifying an individual human being with a computer program appears to be a fundamental tenet of faith among some of the AI community, and lends support to the view that scientists somehow lack humanity.

Both art and science are needed. It is a misunderstanding to suppose them to be mutually antagonistic rather than complementary. No doubt science and poetry are very different. Science strives to write sentences that are unambiguous as regards content, largely paraphrasable and infinitely translatable even into languages that have very different tempi and nuances. At its inception the Royal Society explicitly set its face, quite properly, against metaphor and rhetoric in general. Poetry, on the other hand, emerges with the form, style and choice of word that evokes a unique response in the reader. Poet, poem and reader form a kind of resonant entity by which thoughts, moods, emotions could not otherwise be celebrated, there being insufficient vocabulary in all the world to describe all delicate shades of meaning. A poem may be paraphrasable, but its essence is thereby lost. That goes for any work of art.

It seems that our understanding of the universe, in the broadest sense, requires the complementary approaches of poetry (art in general) and science. As in quantum theory, a Principle of Complementarity prevails. In quantum theory the physical quantities describing a particle, position and momentum are not mutually independent as they are in classical physics. Both quantities can be measured with arbitrary precision in classical physics, but in the quantum world the accurate measure of one precludes the accurate measure of the other – focusing on one blurs the other. In an analogous way, a classical view would see poetry and science as mutually independent and each applicable universally. In such a view the poetic description of mechanical interactions would be meaningful, and the Naturalistic Fallacy would not, in fact, be a fallacy, all of which is nonsense. A view borrowed from quantum theory, on the other hand, sees poetry and science as complementary, the regime of applicability of one precluding the other. In any context, the more poetry, the less science, the more science, the less poetry; but both are needed if our understanding is to span all that there is.

Art and science were once part of the same magico-religious tradition. The crystallization of science out of the heady brew of astrology and alchemy helped as much to define what magic was as what science was. The magic purporting to control the physical world passed over into science; that purporting to control demons and spirits was sanitized within religion or went sheepishly underground. Magic remained embodied in the power of words, of music, of pictures and symbols, of personality, informing art to the present day. The creative, imaginative force of the old tradition is as strong as ever, but now manifested more coherently in art and science. It continues to fuel our desire for understanding, and we ought to appreciate the complementary contributions that art and science make, and be intensely interested in what the best of both have to say.

ONE   INTRODUCTION

1 C. P. Snow, The Two Cultures: A Second Look (Cambridge, 1965).
2 Mary Midgley, Science as Salvation (London: Routledge, 1992).
3 John Horgan, The End of Science (New York: Addison-Wesley, 1997). See also David Lindley, The End of Physics (New York: Basic Books, 1993).
4 Richard Dawkins, The Selfish Gene (Oxford, 1976).
5 John Ziman, Reliable Knowledge (Cambridge, 1978).

TWO   THE LIMITS OF SCIENCE

1 David Hume, A Treatise of Human Nature (Glasgow: Fontana, Collins, 1962).
2 Immanuel Kant, Critique of Pure Reason (Buffalo, New York: Prometheus Books, 1990).
3 Edward O. Wilson, Consilience (New York: Knopf, 1998).
4 Erwin Schrödinger, What is Life? Mind and Matter (Cambridge, 1980).
5 Richard Dawkins, The Selfish Gene (Oxford, 1976).
6 Richard Dawkins, The Selfish Gene (2nd edn, Oxford, 1989).
7 R. G. Collingwood, The Principles of Art (Oxford, 1958). See also Aaron Ridley, R. G. Collingwood (London: Phoenix, Orion, 1998).
8 Stephen Hawking, 'Is the end in sight for theoretical physics?' Physics Bulletin 32 (1981): 15.
9 Kurt Gödel, On Formally Undecidable Propositions of Principia Mathematica and Related Systems (see, for example, New York: Dover Publications, 1992).
10 A readable account is found in Andrew Hodge, Alan Turing, The Enigma (London: Vintage, 1983).

THREE   METASCIENCE

1 Karl R. Popper, *The Logic of Scientific Discovery* (London: Hutchinson, 1977).
2 Paul K. Feyerabend, *Against Method* (London: Verso, 1978).
3 See, for example, John D. Barrow, *Theories of Everything* (New York: Fawcett Columbine, 1991).
4 Richard P. Feynman, Robert B. Leighton and Matthew Sands, *The Feynman Lectures on Physics* (Reading, Mass.: Addison-Wesley, 1963).
5 Benedictus de Spinoza, *Ethics* (London: Dent, 1723).
6 See, for example, Frederick G. Weiss, *Hegel, The Essential Writings* (New York: Harper and Row, 1974).
7 Alfred North Whitehead, *Science and the Modern World* (London: Penguin, 1938).
8 John D. Barrow and Frank J. Tipler, *The Anthropologic Cosmological Principle* (Oxford, 1986).
9 See, for example, Albert Einstein, *Ideas and Opinions* (New York: Dell, 1981).
10 ibid.
11 Alan Guth, *The Inflationary Universe* (Reading, Mass.: Perseus Books, 1997).

FOUR   SCIENCE AND MAGIC

1 The literature on Renaissance magic is extensive. Most of the material in this chapter derives from the books by Frances A. Yates, for example, *Giordano Bruno and the Hermetic Tradition* (London: Routledge & Kegan Paul, 1964) and *The Rosicrucian Enlightenment* (London: Routledge & Kegan Paul, 1972). See also Richard Cavendish, *The History of Magic* (London: Arkana, Penguin, 1987).
2 Francis Bacon, *The Advancement of Learning* (1605).
3 Arthur O. Lovejoy, *The Great Chain of Being* (Cambridge, Mass.: Harvard University Press, 1936).
4 Henri Bergson, *Mind Energy* (London: Macmillan, 1920).
5 Ilya Prigogine, *From Being to Becoming* (San Francisco: W. H. Freeman, 1980).
6 D. J. Bohm and B. J. Hiley, *The Undivided Universe* (London: Routledge, 1993).
7 Francis Bacon, *The New Atlantis* (1627).
8 William Gilbert, *De Magnete* (1600).

SIX   THE MUSIC OF THE SPHERES

1 J. L. E. Dreyer, *A History of Astronomy from Thales to Kepler* (New York: Dover, 1953).

SEVEN   SCIENCE AND MATHEMATICS

1 Francis Bacon, *Novum Organum* (1620).
2 René Descartes, *Rules for the Direction of the Mind* (1628).
3 Blaise Pascal, *Pensées* (1669).
4 I have found the following books extremely worth while:
   G. H. Hardy, *A Mathematician's Apology* (Cambridge, 1940).
   Tobias Dantzig, *Number, The Language of Science* (New York: Macmillan, 1954).
   Morris Kline, *Mathematics, The Loss of Certainty* (Oxford, 1980).

EIGHT   NUMBERS

1 Sir Arthur Eddington, *Fundamental Theory* (Cambridge, 1953)
2 John D. Barrow and Frank J. Tipler, *The Anthropic Cosmological Principle* (Oxford, 1986).

NINE   QUANTUM MAGIC

1 A comprehensive account of the basic issues in quantum theory can be found in the book by Max Jammer, *The Philosophy of Quantum Mechanics* (New York: John Wiley, 1974).
2 Murray Gell-Mann, *The Quark and the Jaguar* (New York: W. H. Freeman, 1994).
3 David Bohm, *Quantum Theory* (Englewood Cliffs, N.J.: Prentice-Hall, 1951).
4 Roger Penrose, *Shadows of the Mind* (Oxford, 1994).
5 Stephen Hawking and Roger Penrose, *The Nature of Space and Time* (Princeton, 1996).
6 David Bohm and B. J. Hiley, *The Undivided Universe* (London: Routledge, 1993).
7 Bernard d'Espagnat, *Conceptual Foundations of Quantum Mechanics* (Reading, Mass.: W. A. Benjamin, 1976).
8 Roland Omnès, *The Interpretation of Quantum Mechanics* (Princeton, 1994).
9 Andrew Steane, 'Quantum Computing', *Rep. Prog. Phys.* 61 (1998): 117.

TEN   SCIENCE AND THE MIND

1 Sir Charles Sherrington, *Man on his Nature* (London: Penguin, 1955).

2 Roger Penrose, *The Emperor's New Mind* (Oxford, 1989).

3 John R. Searle, *The Mystery of Consciousness* (London: Granta Books, 1997).

4 Roger Penrose, *Shadows of the Mind* (Oxford, 1994).

5 John Horgan, *The End of Science* (New York: Addison-Wesley, 1997).

6 P. W. Anderson, 'Shadows of doubt', *Nature* 372 (1994): 288.

7 Horgan, op. cit.

8 Hilary Putnam, book review, *New York Times*, 20 November 1994.

9 See also the voluminous response to *The Emperor's New Mind* in *Behavioural and Brain Sciences* 13 (1990): 643.

10 P. M. S. Hacker, *Wittgenstein* (London: Phoenix, 1997).

TWELVE   SCIENCE AND ART

1 R. G. Collingwood, *The Principles of Art* (Oxford, 1938).

THIRTEEN   SCIENCE AND SENSIBILITY

1 Francis Bacon, *The Advancement of Learning* (1605).

2 Friedrich Nietzsche, *The Genealogy of Morals* (1887).

3 Frank Tipler, *The Physics of Immortality* (London: Macmillan, 1995).

Printed in the United States
by Baker & Taylor Publisher Services

Printed in the United States
by Baker & Taylor Publisher Services